Basic Statistical Computing

D Cooke

A H Craven

G M Clarke

Mathematics Division, University of Sussex

Edward Arnold

© D Cooke, A H Craven, and G M Clarke 1982

First published 1982 by
Edward Arnold (Publishers) Ltd
41 Bedford Square, London WC1B 3DQ

Reprinted 1982, 1984

British Library Cataloguing in Publication Data

Cooke, D.
  Basic statistical computing.
  1. Mathematical statistics – Data processing
  I. Title   II. Craven, A. H.   III. Clarke, G. M.
  519.5′028′5404   QA276.4

  ISBN 0-7131-3441-0

Typeset by Macmillan India Ltd., Bangalore
Printed and bound by
Thomson Litho Ltd., East Kilbride, Scotland.

Preface

Our aim in writing this book has been to show how a computer can best be used to help analyse statistical data, especially for readers who are not highly expert in programming and will be likely to use a microcomputer (e.g. PET, APPLE, TRS80, RM 380Z) rather than a large mainframe installation. We thus emphasise simple, fundamental procedures of statistics, and for descriptions of these we frequently refer to the text *A Basic Course in Statistics*, by G M Clarke and D Cooke (Edward Arnold, 1978): references are given in the form '*ABC*, Section 0.0'.

The language most commonly available on microcomputers is BASIC; all the programs in this book are written in BASIC. The reader should note that there are some differences in the way BASIC is implemented on different machines, and we note one or two of these in the text. We have taken the specification of BASIC in this book to be that given in *Basic BASIC*, by D M Monro (Edward Arnold, 1978); readers whose machines find difficulty in operating any of our programs may profit from comparing Monro's specification with the manual accompanying their own machine.

The development of programs for statistical problems involves using a wide range of programming techniques, and we consider that teachers of computing could find this text very useful, providing an interesting background to many essential methods. It could well be used for a course in computer programming running in parallel with a basic course in statistics, in sixth forms of schools and also in colleges and universities.

We have included chapters on regression and on analysis of variance, because these are methods needed widely by users of statistics. Thus the text could also serve the needs of the non-specialist user of statistics, e.g. in a research institute that has its own microcomputers: all the commonly-used statistical measures and techniques are covered.

A few sections and exercises are rather harder than others: they have been marked *. In Appendix C, there is a large set of data, parts of which can be used in various exercises throughout the text: suitable exercises are marked†. Additional suggestions for the use of the data are given on pages 149–50.

We are grateful to Mr I A Potts, of Ardingly College, and to our colleagues Dr A J Weir and Mr W A Craven, who have helped us test algorithms on various machines, and it is a pleasure once again to thank Mrs Jill Foster for her invaluable assistance in preparing typescript.

Sussex   D Cooke
March 1981   A H Craven
                                                                         G M Clarke

Contents

List of Algorithms and Programs

Notation

Variable names used in the algorithms. Major usage of the names is covered but the list is not exhaustive.

A	intercept in regression equation
A1, A2, ...	coefficients in mathematical expressions
A($\cdot$)	data array for sorting
B	slope of regression line
B($\cdot$)	vector of regression coefficients
C	number of classes
C($\cdot$)	vector of column totals of table
C\$($\cdot$)	array of column descriptors
D1	class-width; strip width in Simpson's rule
E($\cdot$)	vector of expected frequencies
F	F-statistic; value of F random variable
F0	rounding parameter (number of decimal places)
F($\cdot$)	vector of frequencies in classes; vector of fitted values
F($\cdot,\cdot$)	matrix of frequencies
G	gap length in Shellsort
G($\cdot$)	vector of data for Wilcoxon test
H0	window height (an output parameter)
H($\cdot$)	vector of data for Wilcoxon test
I	counter
I1, I2	endpoints of a class
J	counter
J1, J2	labels
K	counter
K0	probability associated with confidence interval
K1, K2	degrees of freedom
L	lower bound (minimum) of variable; Poisson parameter
L0	lower limit of integration (Simpson's rule)
L1, L2	lower bounds of variables 1, 2; lower endpoints of variables 1, 2
M	mean
M0	margin (an output parameter); mean on NH
M1, M2	means of variables 1, 2
N	number of observations; binomial parameter
N0	number of strips in Simpson's rule
N1	number in sample 1; number of treatments; number of rows in table
N2	number in sample 2; number of blocks; number of columns in table

N(·)	vector of number of replicates of treatments
O(·)	vector of observed frequencies
P	parameter of Bernoulli, binomial, geometric distributions; in P-th Q-tile
P∅	integral from Simpson's rule; value of distribution function
P(·)	vector of probabilities of discrete random variable
Q	binomial probability (1–P); in P-th Q-tile
Q∅	value of P-th Q-tile
Q1, Q2, Q3	quartiles
Q(·)	vector of quartiles
R∅	plot residual
R(·)	vector of residuals: vector of ranks; vector of row totals
R(·,·)	matrix of residuals
R$(·)	vector of row descriptors
S1, S2, S3	sums of squares in analysis of variance
S(·,·)	matrix of sums of squares and products
T	total; Wilcoxon statistic
T∅	*t*-statistic
T$	title
T1, T2	totals; rank sums of samples 1, 2
U	upper bound (maximum) of variable
U∅	upper limit of integration (Simpson's rule); Mann–Whitney statistic; uniform pseudo-random variable
U1, U2	upper bounds of variables 1, 2; upper endpoints of variables 1, 2
V	estimated variance
V1, V2	estimated variances of variables 1, 2
W, W1, W2, . . .	working variables
W∅	window width (an output parameter)
W(·)	vector of locations
X	observation; generated random variable
X1	generated random variable
X2	chi-squared statistic; chi-squared value; generated random variable
X(·)	vector of data
X(·,·)	matrix of data
X$	name of variable
Y(·)	vector of data
Z	value of standard normal variable

ln stands always for $\log_e$.

1 Introduction

1.1 The development of calculating devices in statistics

The rapid growth of statistics is a feature of the twentieth century. It has benefited from the development of more powerful calculating machines and computers. By 1900, precision engineering was far enough advanced for efficient calculators to be produced at attractive prices—the eminent statistician Karl Pearson wrote in a letter of 1894: 'I want to purchase a Brunsviga calculating machine before anything else, and am making inquiries about it. I think it would make moment-calculating easy.' Another development of the time was due to Hollerith, in the United States, who wanted to automate the tabulation of census data. He represented data by holes punched in cards, and devised electrical machines to sort and count the punched cards: these were used in the 1890 Census in the USA. Hollerith's company for manufacturing these machines has become part of International Business Machines (IBM), well-known today as computer manufacturers.

In the early 1900s, routine statistical analysis mostly involved tabulating data and calculating averages and index numbers. Simple adding and listing machines then used could add and subtract, print out totals, sub-totals, and individual items if required; Hollerith's punched-card machines did this on a large scale, and were used by banks, insurance companies, business, and government departments. They helped in tabulating data from censuses and large sample surveys. The desk calculator, referred to by Karl Pearson, consisted of circular discs with the digits 0 to 9 marked on them; these were mounted on a common horizontal axis and by turning a handle the machine could be made to add, subtract, multiply and divide. Desk calculators marked a revolution in the development of statistics, comparable to that due to computers in the 1960s.

Since early this century, routine statistical analysis of experimental data has been a part of agricultural and biological research; more recently it has become so in medical and industrial research. These analyses were very largely done on desk calculators, and Fisher (the man who has had most influence on the development of statistics) once said that most of his statistics had been learnt on the machine: without calculators, neither Karl Pearson nor Fisher could have achieved what they did. During these analyses of data, Fisher and his colleagues devised new statistical methods to deal with small samples of data and with data from designed experiments. Means and variances were of particular interest, and the central technique was the analysis of variance (see Chapter 12): this has as its basis a decomposition of a sum of squares. Calculating sums of squares was fairly straightforward on desk calculators, and became particularly easy when (some 40 years ago) the machines became powered by electric motors and incorporated extra keys to speed up the process of summing squares. Another development was to give machines extra storage registers, to hold intermediate steps in a calculation without so much need to copy numbers down on paper: this is often a source of errors. Large desk machines, used by research workers, could carry a considerable number of digits in a calculation; up to twenty in a sum of squares.

The first automatic electronic computers began to appear in the 1940s, and by 1960 many large companies, government departments, research institutes and universities had computer departments based on these machines. The important characteristic of the electronic computer is the very high speed at which it carries out calculations; hence it has to be automatic since no human brain could react quickly enough to keep control of a series of calculations. The computer must therefore be controlled by a program, or sequence of instructions, stored within it.

It was obvious that computers were very valuable in reducing large quantities of data from censuses and surveys, but most statisticians tend to be concerned with much smaller sets of data. Hence statisticians were slow to take up the use of electronic computers (although Rothamsted Experimental Station installed one in the early 1950s). They were put off by the lack of flexibility in analysis when using a computer; the early programs were mostly written by mathematicians and did not do exactly what the statisticians wanted, and the statisticians were on the whole unwilling to devote a lot of time to writing programs so long as their desk machines could do most of what they required. In a computer analysis, one loses direct contact with the data and cannot keep it, and the steps in an analysis, under close scrutiny as a statistician wishes to do—and can easily do when using a desk calculator. Although in theory the computer can carry out various checks during an analysis, early programs were not generally available to do this to the satisfaction of many statisticians; but perhaps the main criticism was of the 'one-shot' nature of the programs. The precise steps needed in a full statistical analysis can rarely be predicted completely at the outset, choice of later computations being very likely to depend on the results of earlier steps. In fact the chief users of statistical programs supplied by computer departments were, and most probably still are, non-statisticians. A research worker who has collected a set of data, very likely without consulting a statistician at the planning stage or even without any very precise plan at all, often sees his analysis as purely a piece of computing. He assumes that the program he chooses—or someone else chooses for him—will automatically give a satisfactory analysis of his data. Far too many analyses are done using programs which are not appropriate to the questions a set of data can answer. Often this is simply because a program is readily available; sometimes it is due to sheer lack of understanding or communication.

Some recent developments have made computers more acceptable to statisticians. Large computers often have multi-access interactive systems, with each of many users working at a terminal so that, provided that the system is not overloaded, he can act as if he alone is using the computer. Programs can be developed relatively quickly, and package programs (e.g. GLIM—General Linear Interactive Modelling) have been developed which allow the user to vary his planned analysis as results appear.

But what we see as a most important development for statistical analysis is the recent arrival of microcomputers. There has been a progressive and quite astounding miniaturisation of electronic computers. The early ones depended on valves; replacing these by transistors led to substantial reduction in size for the same level of power, and the more recent use of very small integrated circuit chips has made available desk computers which are of similar size to the old desk calculators, and are (allowing for inflation) also a similar price. But the computing power of a microcomputer is enormously greater than that of the desk machines; they can be programmed using a high-level language like a large mainframe computer, but they do not have the same problems of input and output, the complexity of which can waste much time and cause frustration to the ordinary user. So once again it becomes possible to keep an analysis under the user's control, to keep close to the original data while analysing it, and to watch the successive steps of the analysis for odd and interesting features. It is these considerations that have been foremost in our minds as we wrote this book.

We must not close this section without mentioning the pocket-size hand-held electronic calculators that first appeared in the 1970s. These have much to offer the occasional user of statistics: they allow quite complicated calculations to be done quickly and silently, without having to go to a computer room to punch cards or tape. Like the old electro-mechanical desk machines, they are of course prone to operator error, although unlike them the hand-held machines have a number of inbuilt programs, ranging from a mean-and-variance routine to (in more powerful versions) routines for regression and evaluating probability densities. It can be useful to have one available even when using a computer.

1.2 Programming a computer for statistics

Software does not have the same dramatic history of progress as does hardware. Nonetheless, the development from machine codes, which were the only languages available for early computers, to the high-level languages available today, such as BASIC and FORTRAN, is a substantial achievement. But even if a high-level language is available, there is still the problem of how best to organise statistical computing.

Chambers (1980) points out three main approaches: (a) single programs; (b) statistical systems, or large package programs; (c) collections of sub-programs or algorithms. The single-program approach was adopted in developing the BMD Biomedical Computer Program at the University of California at Los Angeles. This was the first collection of statistical programs to be made generally available, the first manual being published in 1963; the programs have been very successful and continue, after many revisions, to be used. All the user needs to know is how to input and output data. However, with a collection of this type, combining programs may be awkward or impossible.

The second approach is to develop a statistical system. We can regard this as a very complex program allowing the user to carry out a range of statistical analyses by giving instructions in the special language of the system. Examples are SPSS (Statistical Package for the Social Sciences), developed in the USA, and GENSTAT (A General Statistical Program) and GLIM, both developed in Britain. These systems offer much to a regular user who has a full knowledge of the language of the system and appreciates the package's strengths and weaknesses; but the occasional user can find them quite incomprehensible! Usually such a system can be updated only by a complete revision; and since revisions are only likely to take place at considerable intervals of time, the system can easily fossilise.

The third approach is to develop a collection of statistical sub-programs or algorithms, and to combine these into programs as required. A convenient way of combining algorithms must be found, but if this can be done there is great flexibility in this method, which can be updated by introducing new algorithms. This book is essentially a collection of algorithms.

So far the computer has had a relatively small effect on basic statistics such as we chiefly consider in this book. Many of our programs are for procedures which can be done quite well on desk calculators, though of course if a satisfactory program is available the use of a computer reduces the tedium of this task. One vast area opened up by computers is that of simulation (see Chapters 7 and 9). Also the calculation of 'residuals', trivial on a computer (see Section 11.3) but prohibitively heavy on a desk calculator, helps to make possible fuller and more appropriate analyses of data. We shall also give (see Chapter 8) programs that allow the user to dispense with statistical tables and to calculate necessary values directly for himself. Graphical methods are an area in which the computer promises much but because we limit ourselves to what is available in standard BASIC our programs (see Chapter 5) can only hint at some of the possibilities.

The effect of computers is much more noticeable in advanced statistics. A common, and relatively simple, example is multiple regression, for which we give a program in Section 11.4 although we do not discuss it in detail. Regressions using a small number of variables and observations were often carried out on desk calculators, but the arithmetic became overwhelming if many variables were used. With a computer it is easy to include a very large number of variables; although unfortunately as the number increases so do the difficulties in interpreting the results, and in view of its wide use by non-statisticians one might confidently say that multiple regression has become the most misused technique of statistics. The whole field of multivariate analysis has been opened up through the advances in computers, but the techniques developed there are based on more complicated mathematics and require considerable computer storage space; so we have not extended the scope of this book to cover that field.

1.3   Representation of numbers in a computer

When using any calculating device, we must consider what errors will be introduced into calculations. In a computer, there is only a limited amount of space in which to store each number, and only the most significant digits in the number will be retained; the actual number of digits is governed by the word length of the particular machine. For example, consider calculating π^2 on a machine which will store six digits. The number $\pi = 3.1415926535 \ldots$ would be stored as 0.314159×10^1 (the decimal part is called the mantissa and the index of 10 the exponent). Multiplication of π by π would give, at best, $0.098695877281 \times 10^2$, which would be stored as 0.09869×10^2 and returned as 9.8696. As compared with the true value of π^2, this calculation gives an error in excess of 0.000004.

A second source of error is that computers operate on binary, not decimal, numbers. A number is represented in the computer by a sequence of switches, each of which may be *on* or *off*, in this way representing the binary digits 0 and 1. Thus each number used is converted from decimal to binary representation, and when the arithmetical operations are finished the resulting number is converted back to decimal. Whole numbers in the decimal system have exact binary equivalents, but most other real numbers do not, even though they may be decimal fractions. For example, $0.1 = \frac{1}{10}$ has the infinite binary representation 0.0 0011 0011 0011 ... and cannot therefore be stored exactly in the computer. The effect of this is easily seen by running the following short program:

```
10    LET T = 0.0
20    LET T = T + 0.1
30    PRINT T
40    GO TO 20
```

Since most computers convert a decimal number into a decimal fraction with an exponent, as a first step in calculation, in the way that we wrote π as 0.314159×10^1 above, there is almost certainly an extra error involved in converting the mantissa to a binary number. Similar errors occur with most arithmetic operations carried out in a program, and there is a real danger that the error will accumulate to the extent that several significant digits will be in error. In the worst cases, accumulation of round-off error and loss of significant digits can invalidate a calculation completely.

Consider calculating $f(x) = \dfrac{p}{1 - x^2}$ for $p = 3.0$ and $x = 0.99900050$. Because of rounding, x is taken as 0.99900000 (an error in the seventh decimal place). The result obtained is 1500.7504, but it should be 1501.5007: an error in the *seventh* decimal place has produced an error in the *fourth* significant figure of $f(x)$. The critical step here is subtracting x^2 from 1; these numbers are very similar in magnitude, and this leads to significant digits being lost in the divisor. The detailed analysis of such errors is beyond the scope of this book and we recommend the interested reader to consult texts on numerical analysis, e.g. those by Johnson and Riess (1977) and Wilkinson (1963).

1.4 Interactive and batch computing

If we sit at the keyboard of a microcomputer, using the language BASIC to operate it and watching the output on the video-monitor, we are in complete control of the machine and can alter any instruction at will and change data as we wish. We can interact with the program being implemented, and so we call this way of using a machine **interactive computing**.

Alternatively we may have a terminal connected to a much larger machine, and we shall be sharing its facilities with other users. However, fully interactive computing is still possible because the machine shares its time between all the users. Because of the great speed of a modern computer, even though the machine is servicing tasks input from many terminals any particular user experiences only momentary delay and may well feel as though he has individual use of the machine.

It is extremely useful to work in this way when developing programs and processing small amounts of data. But when the sets of data are large, or several analyses have to be carried out, then it may be more convenient, and cause less delay to other users, to offer the program and data to the machine to be stored for eventual processing when interactive facilities are less in demand. This is **batch processing**, and is recommended when using large statistical package programs.

1.5 Exercises

1 Understandability of programs is more important than speed. But if speed can be obtained without a loss of understanding this is very desirable. Compare the times taken by the different codings below; this will indicate some desirable programming practices.

(1) *Repeated evaluation of an expression*, especially inside a loop, can be very wasteful of time; it is better to do the evaluation once (outside the loop if there is one). Compare the times taken to run the following algorithms.

```
(a)  10    LET A = 3
     20        FOR I = 1 TO 2000
     30        LET X = 2*A*I
     40        NEXT I

(b)  10    LET A = 6
     20        FOR I = 1 TO 2000
     30        LET X = A*I
     40        NEXT I
```

Expressions involving constants are evaluated more slowly than those involving variables only: thus (c) will be slower than (b).

```
(c)  10    FOR I = 1 TO 2000
     20    LET X = 6*I
     30    NEXT I
```

It is also likely to be slower to look up an array variable than a simple variable. Compare the speeds of the algorithms (d) and (e).

```
(d)  10    LET X(3) = 4
     20        FOR I = 1 TO 2000
     30        LET X = I*X(3)
     40        NEXT I

(e)  10    LET X(3) = 4
     20    LET A = X(3)
     30        FOR I = 1 TO 2000
     40        LET X = I*A
     50        NEXT I
```

(2) *Multiplication and division* are slower operations than addition and subtraction. Compare the speed of (f), whose central calculation involves two multiplications and a subtraction, with (g) which involves one multiplication and one subtraction.

```
(f)  10    LET A = 3
     20        FOR I = 1 TO 2000
     30        LET X = I*I - A*I
     40        NEXT I

(g)  10    LET A = 3
     20        FOR I = 1 TO 2000
     30        LET X = (I - A)*I
     40        NEXT I
```

(3) *Exponentiation* is very much slower than an equivalent multiplication. Compare the speeds of (h) and (j).

```
(h)  10    FOR I = 1 TO 2000
     20    LET X = I↑2
     30    NEXT I

(j)  10    FOR I = 1 TO 2000
     20    LET X = I*I
     30    NEXT I
```

[Running these algorithms on an APPLE computer gave these times (in seconds): (a) 14; (b) 9; (c) 10; (d) 14; (e) 9; (f) 16; (g) 13; (h) 104(!); (j) 9.]

2 A computer works in four-significant-digit arithmetic. Find the values of ab/c that it will produce when $a = 0.6240 \times 10^4$, $b = 0.2326 \times 10^{-3}$, $c = 0.1918 \times 10^1$, and the calculation is carried out as (i) $[a.b]/c$, (ii) $a.[b/c]$; the brackets [ ] denote which operation is done first.

3 Working in six-significant-figure arithmetic, find the smaller root of

$$0.001 x^2 + 100.1 x + 10000 = 0$$

(a) using the usual formula; (b) using the fact that the product of the roots is equal to the constant term in the equation, divided by the coefficient of x^2.

4 The integral

$$I_n = \frac{1}{e} \int_0^1 x^n e^x \, dx \qquad \text{where } n = 1, 2, \ldots$$

can be calculated from the following recurrence relation (derived by integrating by parts):

$$I_n = 1 - n I_{n-1} \qquad n \geq 1$$

and $I_0 = 1 - \dfrac{1}{e}$.

From the form of the integrand it is clear that $I_n \to 0$ as $n \to \infty$. Investigate what happens if the integral is evaluated by this method, using successively cruder estimates for the value of e; to 9 figures, e = 2.71828183.

2 Some general principles

2.1 Computation of a mean

We shall begin our consideration of statistical program writing by looking at what is probably the most common calculation in statistics: finding the arithmetic mean of a set of data. For a small set of data we do not need a computer to do this. However, if we consider how best to program this simple calculation we can identify some useful general principles of programming, without being distracted by the detail of a complex central calculation.

Suppose you sit down at a computer keyboard and try to produce a program to calculate a mean. You might produce the program that follows. Comments on the steps of the program are given on the right; a typical output is also given.

```
10    INPUT N                    Number of observations
20    LET T = 0                  Set T, the current total, to zero
30    FOR I = 1 TO N             Begin loop
40    INPUT X                    Input an observation
50    LET T = T + X              Add observation to T
60    NEXT I                     Repeat loop for all N items
70    LET M = T/N                Calculate mean
80    PRINT "MEAN = "; M         Print result
90    END
RUN                              To start the run
?5                               ? is a prompt to input data
?7.1                             The five observations are input in turn
?7.8
?8.1
?7.7
?7.6
MEAN = 7.66
```

This program would give us the answers we want. It is using the computer as if it were a desk calculator, except that the loop saves us from having to press the addition key each time we input an item of data and we are able to print the word 'MEAN' to indicate what has been calculated. Are we making full use of the potentialities of the computer, and have we taken account of the snags that are introduced by using a computer to do the calculations? How can we improve the program?

If you use this program for a large set of data or for many sets of data, it is almost certain that, at some time, you will make a mistake when typing a number. You might type in 34 instead of 43 or 17.3 instead of 1.73. You might realise the mistake as soon as you had made it or you might not spot it until you come to check the numbers. Either way it is inconvenient, and it may be impossible to modify the calculation. With a more complex calculation a similar mistake in input might well mean that the whole set of data has to be retyped.

We can overcome this difficulty by replacing each INPUT instruction in the program by a READ instruction which refers to a DATA statement. This is essentially, of course, the method we would use to input the data if we were operating in batch mode. The DATA statements become an integral part of the program and are most conveniently collected together at the end of the instructions. It is easy to make a visual check of the data before the program is run and any mistakes may be corrected using the editing facilities of the computer. Checking is easier if the data are arranged in a regular manner with, for example, the same number of observations in each row and numbers arranged

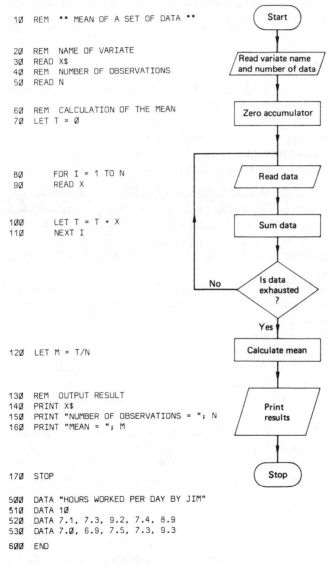

```
10   REM  ** MEAN OF A SET OF DATA **

20   REM  NAME OF VARIATE
30   READ X$
40   REM  NUMBER OF OBSERVATIONS
50   READ N

60   REM  CALCULATION OF THE MEAN
70   LET T = Ø


80       FOR I = 1 TO N
90       READ X


100      LET T = T + X
110      NEXT I


120  LET M = T/N


130  REM  OUTPUT RESULT
140  PRINT X$
150  PRINT "NUMBER OF OBSERVATIONS = "; N
160  PRINT "MEAN = "; M


170  STOP

500  DATA "HOURS WORKED PER DAY BY JIM"
510  DATA 10
520  DATA 7.1, 7.3, 9.2, 7.4, 8.9
530  DATA 7.Ø, 6.9, 7.5, 7.3, 9.3

600  END
```

Program 2.1 *Mean of a Set of Data*

underneath each other. The data can easily be changed for subsequent runs using the editing facilities. We give Program 2.1 in which INPUT has been replaced by READ. The data are introduced in lines 500–530; line 510 indicates the number of days upon which an observation was made, and the lines 520 and 530 give the number of hours on each working day of two successive weeks. The program also incorporates other desirable features, but we present the program first, before discussing these features in the next section.

The input method used in Program 2.1, *Mean of a Set of Data*, allows adequate checking of small sets of data; but for larger sets it is best to use an input algorithm whereby data can be checked, and if necessary modified, as they are entered. An example of this type of algorithm is given in Section 4.2.

2.2   Making a program understandable

We have made other modifications to the program for calculating a mean, in addition to changing the method of input. A program should not only fulfil its purpose efficiently, it should also be easily understandable by another user. (That other user may be yourself in a few months' time—you may well have entirely forgotten or, worse, remember imprecisely what your program will and will not do.) We recommend a number of practices to assist you in making a program intelligible.

(a)   *Give the program a title.* A REMARK statement (REM) at the beginning of a program can contain this title. In this book we have put titles of programs and algorithms between double asterisks, to make them stand out. The computer ignores any line prefaced by REM when the program is being executed. When a program is used repeatedly, modifications are often made; it can be useful to recognise Mark 1, Mark 2, etc., versions of the program.

(b)   *Annotate the program.* Program 2.1 is really too short to require explanatory notes within it. Nevertheless, it is good practice to include such notes wherever they can help to make clear what particular sections of the program are doing or have done. Again we use REM to preface such notes.

(c)   *Choice of variable names.* We have used M for the mean, N for the number of observations and T for the total of the observations. If the name of each variable reminds you what each is doing, this makes the program very much easier to understand. We have chosen particular variable names for the common statistical variables with this in mind and use them throughout this book. A list is given on page xi. BASIC is more restricted than many other high-level languages in the length and style of the names available for variables, so that there is often not as much choice as one would like. For example, we cannot use the name MEAN: we have to be satisfied with M, or M followed by a digit, e.g. M1. We stress that the choice of a variable name should be made with deliberation.

(d)   *Writing the program to exhibit its structure.* It is possible to leave blank lines between sections of the program and to indent loops as we have done in Program 2.1 (although unfortunately these will be lost when the program is reproduced by a computer). This spacing, when coupled with appropriate 'headings' prefaced by REM, has the effect of making the program read rather like the paragraphs in a chapter of a book. Programs should have a definite structure; we return to this point again in Section 2.4.

(e) *Use flow diagrams where appropriate.* Flow diagrams can be useful in planning a program, in tracing faults and in describing the program for another user.

2.3 Making the output understandable

If Program 2.1 were run in batch mode and you obtained the printed output 'MEAN = 7.79', you would probably remember what data had been used on the day you ran it, but would you remember in two weeks' time? You would be in an even greater quandary if several sets of data had been used (the program can easily be modified to do this) and your output read

```
MEAN = 7.79
MEAN = 2.61
MEAN = 5.17
```

The aim should be to make the output self-explanatory even though it adds extra lines to the program. We recommend that the name of a variate, as well as variate values, should be read by the program. This can be done using a string variable, as in line 30 of Program 2.1 which says READ A$, the $ sign indicating a string variable as opposed to a numerical variable. This instruction reads the variate name given in the DATA statement in line 500. String variables must be enclosed in inverted commas. If the variate name is stored in the computer, we can print it at any convenient place in the output and hence clearly label the results. Program 2.1 would produce the output

```
HOURS WORKED PER DAY BY JIM
NUMBER OF OBSERVATIONS = 10
MEAN = 7.79
```

The mean in the output of Program 2.1 is 7.79. The mean of a set of data is more accurate than the individual observations and therefore may be expressed to more decimal places; yet, when the number of observations is small, to quote the mean to many more decimal places would imply that it is more accurate than is really the case. Our present result is, in respect of the number of decimal places quoted, more a lucky chance than something we can rely upon; it has happened simply because we had ten observations and trailing zeros are not usually printed. If we had used a data set with the same total, 77.9, but containing only nine observations, our computer would have given us a value of 8.65556 for the mean. This certainly suggests a greater accuracy than is justified; a more appropriate value would be 8.66. The number of digits printed is determined by a set of rules (e.g. integers are printed exactly, non-integers are presented to six significant figures with trailing zeros suppressed) which takes no account of statistical accuracy. Usually the numbers will be printed to a greater accuracy than is justified, in which case the number of digits should be curtailed if the value is to be quoted; this can be done by using a defined function to round the data (see Section 4.4.5). In some implementations of BASIC it is possible to format the output to give a pre-determined number of decimal places.

2.4   Program structure

A good principle to follow in order to make a program easy to understand and easy to modify is to construct it in distinct sections. If the instructions in each section are preceded by descriptive remarks, the program can, to a certain extent, be its own flow diagram. It is often possible to develop and test sections of the program independently and hence isolate errors more easily. If changes are to be made it is much easier to see what needs to be done if one's efforts can be concentrated on a small section instead of the entire program.

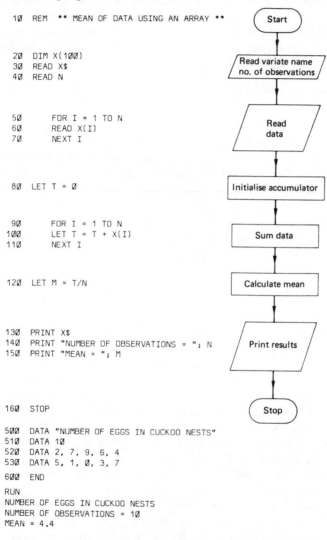

```
10   REM  ** MEAN OF DATA USING AN ARRAY **

20   DIM X(100)
30   READ X$
40   READ N

50       FOR I = 1 TO N
60       READ X(I)
70       NEXT I

80   LET T = 0

90       FOR I = 1 TO N
100      LET T = T + X(I)
110      NEXT I

120  LET M = T/N

130  PRINT X$
140  PRINT "NUMBER OF OBSERVATIONS = "; N
150  PRINT "MEAN = "; M

160  STOP
500  DATA "NUMBER OF EGGS IN CUCKOO NESTS"
510  DATA 10
520  DATA 2, 7, 9, 6, 4
530  DATA 5, 1, 0, 3, 7
600  END
RUN
NUMBER OF EGGS IN CUCKOO NESTS
NUMBER OF OBSERVATIONS = 10
MEAN = 4.4
```

Program 2.2 *Mean of Data using an Array*

   Program 2.1 is quite satisfactory for the simple job it has to do, but it is usually a good general rule to split a program into three main parts which may be described by

        READ
        CALCULATE
        PRINT

Each of these main sections can be subdivided if the nature of the problem demands it. To obtain this simple structure for Program 2.2, *Mean of Data using an Array*, we have to introduce an array X in which the observations are stored. Note that we have to include, in line 20, a DIMENSION statement for the array X; we have allowed for a maximum of 100 observations.

We pay a price, in this case, for this simplification of the program. Extra storage space is required for the array; also the addition of an extra FOR loop and the need to search an array means that the program uses marginally more computer time. Having the data in an array would allow us to extract sections of the data if we wished, though it hardly seems likely that we would need to do so in the present context. If we wish to recover the complete set of data we can do so with a RESTORE statement; an array is not then necessary.

2.4.1 Algorithms

We recommend a modular approach to program construction. For many of the statistical procedures that we consider, we shall find it convenient to list only an appropriate module, or subprogram, rather than a complete program. For example, in Program 2.2, the subprogram that calculates the mean is listed in lines 80–120; the addition of an input section in lines 20–70 and an output section in lines 130–160 combine to form a complete program. We shall call a subprogram, such as that for calculating a mean, an **algorithm**; it is a recipe for carrying out a particular task. Input and output sections (we could call these algorithms as well) must always be added to a calculation algorithm in order to produce a complete program. But a program may consist of many algorithms strung together. If an algorithm is to be used many times in a program we would probably make use of the facilities in BASIC for declaring subprograms as **subroutines** or, in the case of one-line algorithms, as user-defined **functions**.

We have adopted a standard method of writing out algorithms; full details of this are given in Appendix A, which the reader should refer to as necessary. The most important convention to note is that each algorithm includes:

(a) remarks at the beginning to specify the variables needed as input to the algorithm;
(b) remarks at the end to specify the variables which are calculated by the algorithm.

Appendix A also discusses how to make an algorithm into a program, and how to combine many algorithms into a program.

2.4.2 Properties of a good program

The prime aim of a good program is to carry out its task successfully and accurately. In addition, if people are to use the program well, the listed program and the output must have a clear structure and be easy to understand. These properties are the ones that we shall stress in this book.

It is also desirable that a program be efficient in its use of storage, and be fast. We discuss efficiency when considering the relative merits of alternative programs, and we believe that the programs we are recommending are efficient. But, for readers of this book, whom we assume to be mainly inexperienced programmers, we shall regard speed in running a program as less important than success, accuracy and understandability of the program itself.

2.5   Exercises

1   Run Program 2.2, *Mean of Data using an Array*, with the set of data given in lines 510–530 of the program, and check that your result agrees with that given.

2   Modify Program 2.2 so that it can deal with several sets of data, and print out the mean for each set, when used (a) interactively; (b) as a batch procedure.

3   Data on the performance of a new machine are collected over a period of several days; each day several records of a quantity X are taken. At the end of each day, it is required to find (i) M_1, the mean of that day's observations, and (ii) M_2, the mean of all the data collected so far. Write a program which does this *without* working through every item of previous days' data, but updates the previous M_2 instead.

4   Modify Program 2.2 so that it calculates and prints the geometric mean as well as the arithmetic mean. [The geometric mean of a set of numbers $x_1, x_2, \ldots, x_n$ equals $\sqrt[n]{(x_1 x_2 \ldots x_n)}$; it may be calculated by forming the product and taking the nth root, although here there is a danger that the product may overflow the register, or by using the fact that

$$\text{LOG (GEOMETRIC MEAN)} = \frac{1}{n} \sum_{i=1}^{n} \ln x_i.$$

Try both methods of calculation. As a test set of data for x, use the values 1.0 E2, 1.0 E3, 1.0 E4, 1.0 E5, 1.0 E6, which have geometric mean 1.0 E4 $= 10^4$.]

5   It may be inconvenient to have to input the number of observations before each set of data. One way of overcoming this, which can be useful if applied carefully, is to put a **marker**, which is a very large number like 999999, at the end of each data set and to incorporate in the input program a test of when the end of the set is reached. (A disadvantage of this is that 'outliers', or members of the data set which are large, although not as large as the marker, might also need to be traced for study.) A negative integer can be used as a marker if the data themselves are certain to be positive.

Modify Program 2.2 to contain a marker and an appropriate test.

3 Sorting and ranking

3.1 Introduction

A number of useful descriptive statistics and statistical tests depend on relative magnitudes, or order properties, of observations. For example, we may wish to pick out the maximum and minimum observations in a set, or the middle observation (the median) in a set. Sometimes it may even be useful to list a whole population or sample in order of size, from smallest to largest. Several statistical tests use the **ranks**, i.e. the position numbers of the observations when the whole set has been put in order.

The fundamental procedure used in determining most of these quantities is **sorting** the data. Sorting is an extremely important computer method, and much has been written about it. We have space to touch only the borders of the subject; thorough discussions are given in Knuth (1973) and Lorin (1975).

3.2 The Exchange sort

The most efficient methods of sorting are not easy to understand. So we shall give first a method that is easy to understand and simple to program; it is satisfactory to use for small collections of numbers—say less than 30—but very slow for large collections.

We illustrate the method with a small set of data, given in the order

   10, 3, 14, 17, 2, 11.

Begin with the observation on the extreme left, 10; compare it with the observations to the right. If any of these are smaller, as are 3 and 2, choose the smallest, 2, and exchange it with 10:

   2, 3, 14, 17, 10, 11.

Now consider the second observation from the left, 3; compare it with the observations to its right. None is smaller, so no exchange is made. In the third position is 14; we find that the smallest number to its right is 10, and so we exchange these two:

   2, 3, 10, 17, 14, 11.

It is easy to see that by following this procedure we end with the sequence:

   2, 3, 10, 11, 14, 17.

In Algorithm 3.1, *Exchange Sort*, the data are assumed to be in an array $A(\cdot)$. An array name different from the usual $X(\cdot)$ is chosen because in most sorting methods the observations are moved and we may want to preserve the original data order also, in $X(\cdot)$. The array $X(\cdot)$ can be read into $A(\cdot)$ before beginning to sort.

```
500   REM  ** EXCHANGE SORT **
510   REM  * INPUT: DATA A(·), NUMBER OF DATA N *
520   REM  * VARIABLES: I, J, J1, W *

530       FOR I = 1 TO N - 1
540       LET J1 = I
550           FOR J = I + 1 TO N            Compare ith observation
560           IF A(J) >= A(J1) THEN 580     with those to the right
570           LET J1 = J
580           NEXT J
590       IF J1 = I THEN 630

600       LET W = A(I)                      Exchange observations
610       LET A(I) = A(J1)
620       LET A(J1) = W
630       NEXT I

640   REM  * OUTPUT: DATA IN INCREASING ORDER IN A(·) *
```

Algorithm 3.1 *Exchange Sort*

3.3 Shellsort

With N items of data, the number of comparisons made in *Exchange Sort* increases approximately as N^2. The time taken to sort a small set of data is quite acceptable, and the simplicity of the algorithm makes it a good choice. But if it is often required to sort large sets of data, *Exchange Sort* may be irritatingly slow for routine use. It becomes worthwhile to use a more efficient method, even at the expense of more complex programming.

We give next a method, first proposed by D L Shell, that has become generally known as Shellsort; it is also called the 'merge-exchange sort'. For large sets of data it is appreciably quicker than *Exchange Sort*: the time taken increases approximately as $N \ln N$. We illustrate the method by sorting the twelve observations:

$$7, \ 9, \ 2, \ 11, \ 4, \ 1, \ 3, \ 10, \ 5, \ 6, \ 2, \ 8.$$

Begin by comparing all pairs of observations which are $N/2$ apart (or, more strictly, to cover cases where N is not even, $\text{INT}\,(N/2)$ apart); exchange the observations in a pair if they are not in the correct order already. For our set of twelve observations, we list on the same line the pairs to be compared and underline the observations which are exchanged.

First pass: 7   3
               9   10
               2   5
               11   6
               4   2
               1   8.

Having made the exchanges, the data are next merged:

    3, 9, 2, 6, 2, 1, 7, 10, 5, 11, 4, 8.

The gap between observations to be compared is now halved (working if necessary to the next integer below); in our example the gap becomes 3. We list again on the same line the sets to be compared, and underline those observations which are moved.

Second Pass:

3	6	7	11
<u>9</u>	<u>2</u>	<u>10</u>	<u>4</u>
<u>2</u>	<u>1</u>	5	8.

Let us consider the exchanges and comparisons involving the numbers on the second line (these operations take place between operations on other numbers). First, the number '9' is compared with the number '2' and an exchange made:

    2   9   10   4.

Then '9' is compared with '10', and these numbers remain as they are. Next '10' is compared with '4' and these two numbers are exchanged:

    2   9   4   10.

When it makes sense to do so, a primary comparison is followed by a number of secondary comparisons to ensure that all the numbers that have been inspected on a line are in order. Thus the primary comparison, and exchange, of '10' with '4' is followed by the secondary comparison, and exchange, of '4' with '9':

    2   4   9   10;

it is followed also by the secondary comparison of '4' with '2', although this leads to no exchange. The observations are again merged:

    3, 2, 1, 6, 4, 2, 7, 9, 5, 11, 10, 8.

The gap between observations to be compared is next reduced to INT $(3/2)$, i.e. 1, which will be the final gap length every time Shellsort is used.

Third Pass: 3 2 1 6 4 2 7 9 5 11 10 8.

In this example, all observations move in the final pass, as a consequence either of a primary comparison or of a secondary comparison. But because of the earlier long-distance comparisons they move little in the final pass.

Final Merge: 1, 2, 2, 3, 4, 5, 6, 7, 8, 9, 10, 11.

In Algorithm 3.2, *Shellsort*, the gap length is denoted by G. Note that, in line 590, observation J is compared with observation $J + G$ in the original list. If the condition fails, which means that the two observations are already in increasing order, the program passes out of the loop on J (through lines 600, 610 and 650) and the next pair of observations is considered. Otherwise, observations J and $J + G$ are exchanged (in lines 620–640) and as a secondary step observation $J - G$ is compared with observation J; the consequence of this comparison may again be either to pass out of the loop or to make an exchange and stay in the loop.

```
500   REM  ** SHELLSORT **
510   REM  * INPUT: DATA A(·), NUMBER OF DATA N *
520   REM  * VARIABLES: G, I, J, J1, W *

530   LET G = N
540   LET G = INT(G/2)                              Set gap length
550   IF G = Ø   THEN 68Ø
560       FOR I = 1 TO N - G
570           FOR J = I TO 1 STEP - G
580           LET J1 = J + G
590           IF A(J) > A(J1) THEN 62Ø              Compare observations
600           LET J = Ø
610           GO TO 65Ø

620           LET W = A(J)                          Exchange observations
630           LET A(J) = A(J1)
640           LET A(J1) = W

650           NEXT J
660       NEXT I
670   GO TO 54Ø

680   REM  * OUTPUT: DATA IN INCREASING ORDER IN A(·) *
```

Algorithm 3.2 *Shellsort*

3.4   A comparison of sorting methods

In order to show how much the different sorting methods can vary in speed, we have
run a number of sorting algorithms on the same sets of observations; we have recorded
the times taken and other relevant data. Besides *Exchange Sort* and *Shellsort*, we have
included in these trials three further methods. One of these is 'Bubblesort' (or
'Ripplesort'), which is widely used but which is not a method that we would
recommend; the algorithm is described in question 4 of Exercises 3.9. The other two
methods are very rapid; we describe them briefly here and give algorithms for them in
Appendix B.

3.4.1   Quicksort

Algorithm A.1, *Quicksort*, is given on page 140. The complete list of observations to be
sorted is partitioned into two sub-lists, such that the numbers in one sub-list are all less
than or equal to a given number, called the **fence**, while the numbers in the other sub-list
are all greater than the fence. This partitioning process is repeated on the sub-lists, using
fences chosen from each sub-list, and is continued until sub-sub- . . . -lists are produced
containing sufficiently few numbers that it is efficient to sort them by one of the simple
methods, such as *Exchange Sort*. *Quicksort* works by moving numbers in the array A (·)
(as do *Exchange Sort* and *Shellsort*) so that when the final sub- . . . -lists have been
sorted the complete array is in order.

3.4.2 Monkey Puzzle sort

Algorithm A.2, *Monkey Puzzle Sort*, is given on page 141. All the sorting algorithms described so far have involved exchanges within the data array. Since each exchange involves three instructions, a method that could avoid exchanges would have a great advantage. In the Monkey Puzzle sort, each item in the array is considered in turn and, by means of pointers, is located on a branching tree structure (hence the name) in such a way that it is possible by following the pointers to read out the numbers in order. Two additional arrays, $L(\cdot)$ and $R(\cdot)$, each of the same size as the data array, are needed to store the pointer information.

All the sorts may be modified to sort lists of words, or records which may contain several words and numbers. Moving these string variables around is a slow operation, and the Monkey Puzzle sort has the advantage that it does not move the variables. It can also be adapted to sort several variables simultaneously. For example, a set of records might consist, for each member of a group of people, of the person's name, age and address. By including extra arrays of pointers, these variables could be sorted with a single pass through the data.

3.4.3 Sorting times

The sorting times for the five methods we have introduced are given in Table 3.1, which also contains information on the numbers of comparisons and exchanges. There are no

Table 3.1

Time (seconds) to sort n random numbers

n	25	50	100	500
Bubblesort	5.6	22.6	92	2397
Exchange Sort	3.4	13.1	51	1240
Shellsort	3.2	8.6	22	155
Monkey Puzzle Sort	2.9	6.8	16	112
Quicksort	2.8	5.7	12	76

Number of comparisons when sorting n random numbers

n	25	50	100	500
Bubblesort	286	1202	4872	124626
Exchange Sort	300	1225	4950	124750
Shellsort	123	330	827	6006
Monkey Puzzle Sort	215	509	1214	8519
Quicksort	177	435	1050	7329

Number of exchanges when sorting n random numbers

n	25	50	100	500
Bubblesort	147	621	2462	61561
Exchange Sort	21	45	95	492
Shellsort	58	154	396	2916

exchanges in the Monkey Puzzle sort, and the moves in Quicksort cannot be classified as simple exchanges, so there is no figure for 'number of exchanges' given for either of these sorts. The figures given in the tables are averages taken over 100 blocks of *n* random numbers, generated by the PET microcomputer's inbuilt random number generator; for each sorting technique, the generator was initialised identically, so that the blocks of numbers being compared by the various techniques really are the same. Comparisons and exchanges were counted by inserting incremental counters just before each comparison and exchange. The times of execution were found in separate trials using the PET's timing function TI. The instruction LET T1 = TI was inserted as the first executable instruction of the subroutine, and LET T2 = TI was inserted immediately before the RETURN statement. LET T = T2 − T1 thus gave the time spent in the subroutine (in units of $\frac{1}{60}$ sec). The times should be very similar indeed for other machines, although without a timing function it is difficult to time them exactly.

Table 3.1 shows that Bubblesort has little to recommend it. The program is no simpler than that for Exchange sort, and yet it takes almost twice the time. Bubblesort and Exchange sort involve a similar number of comparisons, but Exchange sort requires far fewer exchanges. If we wish to sort sets of about thirty numbers, or less, Exchange sort is a good choice: it is simple to program and, for many purposes, fast enough over this range of sample sizes. When there are many more than thirty numbers to be sorted, Shellsort is much quicker than Exchange sort; the program is not much more complex so that it is well worth using with larger sets of data. In fact the number of exchanges with Shellsort is substantially more than with Exchange sort, but the comparisons are enormously fewer.

The more complex methods of sorting are worth considering only if many large batches of data must be sorted. Of the two methods, Quicksort appears to have the edge over Monkey Puzzle sort in speed when working in BASIC, though it is interesting to note that they both require much the same number of comparisons. Monkey Puzzle sort is expensive on storage; it requires 3*n* locations when sorting *n* numbers, whereas Quicksort only needs $n + \log_2 n$. (Exchange sort and Shellsort each require *n* locations.) Monkey Puzzle sort is useful when there is additional information associated with the items being sorted. In all the sorting methods we have mentioned except Monkey Puzzle, items are moved around while being sorted so that the link with additional information is lost, whereas after Monkey Puzzle sort the additional information is still accessible.

3.5 Ranking

Given an array of data X(·), it may be useful to know the rank of each element in it. These values are calculated in Algorithm 3.3, *Ranking*. The algorithm counts how many observations are smaller than any given observation X(I); this count is labelled T1. If all the observations are different, the rank of X(I) is T1 + 1. Note that the smallest observation has rank 1.

A modification to this simple process is needed if any of the observation-values are repeated. Thus in the ordered set of numbers

10, 11, 13, 14, 14, 17, 19,

```
500   REM   ** RANKING **
510   REM   * INPUT: DATA X(•), NUMBER OF DATA N *
520   REM   * VARIABLES: I, J, T1, T2, W *
530   DIM R(100)

540       FOR I = 1 TO N
550       LET T1 = 0
560       LET T2 = 0
570       LET W = X(I)
580           FOR J = 1 TO N
590           IF X(J) > W THEN 640
600           IF X(J) < W THEN 630
610           LET T2 = T2 + 1
620           GO TO 640
630           LET T1 = T1 + 1
640           NEXT J

650       LET R(I) = T1 + (T2 + 1)/2
660       NEXT I

670   REM   * OUTPUT: RANKS R(•) IN SAME ORDER AS DATA X(•) *
680   REM   * SMALLEST NUMBER HAS RANK 1 *
```

Algorithm 3.3 *Ranking*

there are two 14's, having ranks 4 and 5. In such a case it is the custom to give each occurrence of the repeated number the mean rank, here 4.5; then 17 will be given the rank 6. In Algorithm 3.3, T2 counts how many times each number occurs in the data set. The rank for each occurrence of a number that is repeated is then given as $R(I) = T1 + (T2 + 1)/2$.

This ranking algorithm, 3.3, is satisfactory for small sets of data and has the great virtue of simplicity, but for large data sets it is very slow. It is possible to augment a sorting program to find the ranks of the observations. Algorithm 3.4, *Exchange Sort and Rank*, does this. A location array $W(·)$ keeps track of the movement of the observations. The ranks of the observations, ignoring **ties** (the name given to repeats of the same number), are obtained directly in line 750. The extra lines 770–910 are required to take account of ties.

In a similar way we can augment other sort algorithms in which data are moved in an array; this adds little to the sort time.

3.6 Maximum and minimum values

These values can of course be read off from a data array once it has been sorted; but they can also be found by a single pass through unsorted data, as in Algorithm 3.5, *Maxmin*. The current upper bound, or maximum value, is labelled U and the lower bound, or minimum value, L. Initially both U and L are set equal to the first observation X(1). The values of U and L are modified if necessary as each observation is considered.

```
500   REM  ** EXCHANGE SORT AND RANK **
510   REM  * INPUT: DATA A(•), NUMBER OF DATA N *
520   REM  * VARIABLES: I, J, J1, T1, T2, W, W1 *
530   DIM R(100), W(100)

540   REM  INITIALISE LOCATION ARRAY
550       FOR I = 1 TO N
560       LET W(I) = I
570       NEXT I

580   REM  SORT AND RECORD MOVES
590       FOR I = 1 TO N - 1
600       LET J1 = I
610           FOR J = I + 1 TO N
620           IF A(J) >= A(J1) THEN 640
630           LET J1 = J
640           NEXT J
650       IF J1 = I THEN 720
660       LET W = A(I)
670       LET A(I) = A(J1)
680       LET A(J1) = W
690       LET W1 = W(J1)
700       LET W(J1) = W(I)
710       LET W(I) = W1
720       NEXT I

730   REM  RANKING OBSERVATIONS
740       FOR I = 1 TO N
750       LET R(W(I)) = I
760       NEXT I

770   REM  GIVE EQUAL RANKS TO TIES
780       FOR I = 1 TO N - 1
790       LET T1 = 1
800       LET T2 = R(W(I))
810       IF A(I+1) <> A(I) THEN 870
820       LET T1 = T1 + 1
830       LET T2 = T2 + R(W(I+1))
840       LET I = I + 1
850       IF I = N THEN 870
860       GO TO 810
870       IF T1 = 1 THEN 910
880           FOR J = (I + 1 - T1) TO I
890           LET R(W(J)) = T2/T1
900           NEXT J
910       NEXT I

920   REM  * OUTPUT: RANKS R(•) IN SAME ORDER AS INPUT ARRAY A(•) *
930   REM  * DATA IN INCREASING ORDER IN OUTPUT ARRAY A(•) *
```

Algorithm 3.4 *Exchange Sort and Rank*

```
500   REM   ** MAXMIN **
510   REM   * INPUT: DATA X(•), NUMBER OF DATA N *
520   REM   * VARIABLES: I, L, U, W *

530   LET U = X(1)
540   LET L = U

550       FOR I = 2 TO N
560       LET W = X(I)
570       IF W > U THEN 610
580       IF W >= L THEN 620
590       LET L = W
600       GO TO 620
610       LET U = W
620       NEXT I

630   REM   * OUTPUT: MAXIMUM U, MINIMUM L *
```

Algorithm 3.5 *Maxmin*

3.7   Quantiles

The quantiles are a useful class of descriptive statistics. They are observations (or notional observations) which divide, in specified proportions, the total frequency of a set of observations. The most commonly-used quantile is the **median**: this is the observation that divides the total frequency in half—or, in other words, it is the middle observation. If N (the total frequency) is an odd number, the median is the $(N+1)/2$th observation in rank order: for, assuming that all the observations are distinct, there will be $(N-1)/2$ observations smaller than the median and the same number larger than it. If N is even, there is no 'middle' observation, and so we put the median midway between the $N/2$th and the $(N/2+1)$th observations, since this value will have $N/2$ observations below it and $N/2$ above. This illustrates the common practice when calculating quantiles: if a dividing point does not fall on an observation, we estimate the quantile as the mean of the two observations between which it falls. P should be restricted to the range 1 to $Q-1$ (inclusive) when using Algorithm 3.6.

```
500   REM   ** QUANTILES **
510   REM   * INPUT: DATA A(•) IN INCREASING ORDER *
520   REM   * NUMBER OF DATA N; P, Q FOR P-TH Q-TILE *
530   REM   * VARIABLES: J1, J2, Q0 *

540   LET J1 = INT(P*N/Q) + 1
550   LET J2 = N - INT((Q - P)*N/Q)
560   LET Q0 = (A(J1) + A(J2))/2

570   REM   * OUTPUT: Q0, THE P-TH Q-TILE *
```

Algorithm 3.6 *Quantiles*

The quartiles divide the total frequency into quarters, the deciles divide it into tenths and the percentiles divide it into hundredths. Algorithm 3.6, *Quantiles*, is for calculating the P-th Q-tile from a data array that has already been sorted. Thus $P = 3, Q = 4$ gives the third quartile; $P = 2, Q = 10$ gives the second decile; $P = 1, Q = 2$ gives the median. The quantile is calculated from both ends of the set of data (i.e. beginning with the largest as well as beginning with the smallest); if these two calculated values coincide the quantile coincides with one of the observations. If the values do not coincide, we estimate the quantile to be half the sum of these two values.

3.8 Test data for Algorithms

Algorithm 3.1, *Exchange Sort*. (i) Input: $N = 6$, $A(\cdot) = 2, 7, 1, 8, 2, 8$.
Output: $A(\cdot) = 1, 2, 2, 7, 8, 8$. (ii) See question 1 of Exercises 3.9, below.
Algorithm 3.2, *Shellsort*. Same as for Algorithm 3.1, *Exchange Sort*.
Algorithm 3.3, *Ranking*. Input: $N = 6$, $X(\cdot) = 2, 7, 1, 8, 2, 8$. Output: $R(\cdot) = 2.5, 4, 1, 5.5, 2.5, 5.5$.
Algorithm 3.4, *Exchange Sort and Rank*. Input: as for Algorithm 3.1, *Exchange Sort*.
Output: $A(\cdot)$ and $R(\cdot)$ as in Algorithms 3.1, 3.3.
Algorithm 3.5, *Maxmin*. Input: as given in (i) above for Algorithm 3.3, *Ranking*.
Output: $U = 8$, $L = 1$.
Algorithm 3.6, *Quantiles*. Input: $N = 6$, $A(\cdot) = 1, 2, 2, 7, 8, 8$. $P = 1$, $Q = 2$.
Output: $Q0 = 4.5$.

3.9 Exercises

1 Write a program to sort 25 random numbers using Algorithm 3.1, *Exchange Sort*. The program might run as follows (after a title and any preliminary information).

```
100    LET N = 25
110        FOR I = 1 TO N
120        LET A(I) = RND(1)
130        NEXT I

...    (Sort Algorithm)

1000       FOR I = 1 TO N
1010       PRINT A(I)
1020       NEXT I
1030   STOP
```

2 Vary N in question 1 and record the times taken.

3 Put in increasing order the first 21 digits of π:

> 3.141 592 653 589 793 238 46

4 *Bubblesort*. Comparisons are made between the numbers as in Exchange Sort, but an exchange is made each time a number is found that is out of order. Thus at the end of the first pass the largest number has moved to the end of the list. The next pass, which ends at the last position but one, moves the second largest number to its correct place. An algorithm for Bubblesort, using the same input variables as Algorithm 3.1, *Exchange Sort*, is as follows.

```
530    FOR I = 1 TO N - 1
540        FOR J = 1 TO N - I
550        LET J1 = J + 1
560        IF A(J) <= A(J1) THEN 600
570        LET W = A(J1)
580        LET A(J1) = A(J)
590        LET A(J) = W
600        NEXT J
610    NEXT I
```

How does this algorithm differ from *Exchange Sort*, and why is it slower?

The algorithm used in the comparison of the speeds of sorting methods (Section 3.4) also included the instructions:

```
535    LET L = 0
595    LET L = 1
605    IF L = 0 THEN 620          (Jump to last line of algorithm)
```

This has the effect of finishing the sort as soon as a pass is made in which no exchange takes place.

5 *Although we do not recommend Bubblesort, we include the following for its interest as an exercise in programming technique.*

Bubblesort can be improved by sweeping forwards and backwards in turn. The first sweep (forwards) takes the largest number to the far right; the next sweep (backwards) takes the smallest number to the far left. Modify the program from question 4 above to incorporate this improvement, and compare its speed with that of an unmodified Bubblesort.

6 The time taken in sorting methods should depend on N. Find a relation of the form $t = kN^P$ for Exchange Sort and for Quicksort (see Table 3.1).

7 Modify *Exchange Sort* (Algorithm 3.1) or *Shellsort* (Algorithm 3.2) so that it will put a list of words into alphabetical order. Note that the statement:

```
200    IF A$(I) > A$(J) THEN 300
```

tests whether the string variable A$ (I) is after A$ (J) in alphabetical order, and if it is then a jump occurs to line 300.

Test your program by putting the days of the week in alphabetical order.
[Note: A space is recognised as a string variable, and so must be taken into account when deciding the order of string variables.]

8 A useful method of scanning data to detect 'wild' observations, errors in typing, misplaced decimal points, etc. is to calculate the maximum, minimum and mean values of the data set. It is useful to identify which items in the list are the maximum and the minimum. Develop a program to carry this out using the following procedure.
1) Read name of variate X$.
2) Read the number of data N; read N data into an array X($\cdot$).
3) Find the mean M of the data in X($\cdot$) (see Program 2.2).

4) Find the maximum and minimum values (U, L) of the data (see Algorithm 3.5) and to which items these belong.
5) Print out name of variate, number of data, minimum value, mean and maximum value, with the numbers labelled (e.g. MEAN 6.32) and necessary identifications made.

Run the program with the following set of 10 observations: 13.7, 16.2, 162, 12.9, 11.0, 21.5, 12.6, 14.5, 11.7, 12.1.

9†   Write a program to:
1) calculate the mean M of a set of N data in an array X($\cdot$);
2) calculate the number N_1 of data less than M and the number N_2 greater than M;
3) print the mean M and an index $(N_1 - N_2)/N$.

[This index measures how close the mean is to the median, being 0 if the mean and median coincide and having a large absolute value if they are far apart.]
Run the program using the following two sets of data.
a) Weekly wages, in £, of ten workers (including the manager) in a factory: 96, 95, 160, 94, 95, 93, 98, 92, 97, 91.
b) Heights of ten plants in a greenhouse, recorded in mm: 31, 27, 43, 35, 37, 33, 34, 36, 39, 36.

10*   Amend *Monkey Puzzle Sort* (Algorithm A.1) so that it is able to sort records on two different criteria with a single pass; for example, put names in alphabetical order and ages in increasing order.

11†   Write a program to calculate:
i) the **range** of a set of data, which is the difference between the maximum and minimum observations;
ii) the **interquartile range**, which is the difference between the upper and lower quartiles (one-half of this, the 'semi-interquartile range', is often used instead).

Incorporate this program into one which calculates, and prints the values of, the mean, median, range and semi-interquartile range of a set of observations. Make sure the output is labelled clearly.
Run the program on the following set of data: 87, 67, 98, 57, 74, 100, 83, 60, 99, 88, 54, 72, 78, 75, 93.

12   Write a flow diagram for *Exchange Sort*, Algorithm 3.1.

4 Inspection and summary of data using tables

4.1 Introduction

Tables and diagrams are the main aids available to help us inspect and present data. In this chapter we discuss how the computer can be used to compile tables; in Chapter 5 we shall deal with graphical methods.

Our first concern must always be that the correct data have been put into the computer. For anything more than a small amount of data it is desirable to use an input routine which incorporates a check on the data entered; we discuss this in Section 4.2.

In this chapter, for the first time in the book, we shall illustrate how algorithms may be combined to form a more complex algorithm or a complete program.

The essence of a good table is layout; we therefore also discuss output. It may surprise the reader to learn that input and output are usually much more awkward to program than are calculations. Moreover the facilities in BASIC for input and output are very limited. We often cannot produce as neat and intelligible an output as we would like; but, with a little care, we can produce a substantial improvement on unconsidered, unplanned output.

4.2 Input of an array

Many of the algorithms in this book operate on a data array, usually called $X(\cdot)$. We give an algorithm *Input – Data Array*, 4.1, in which data for an array are entered in small groups so that it is possible to check each group of numbers, and either accept them or modify them before acceptance. A partial run of a program incorporating the algorithm is given immediately after it, to illustrate how it works.

4.3 A six-number summary

The data we shall be considering in this chapter will usually be a sample from some population that we are interested in; occasionally, however, it may be a whole population. For example, we may be given the prices of packets of the same brand of coffee in a sample of shops, or we may have the complete rainfall figures for each day in July in a town in Great Britain.

The first, very simple, summary table we shall look at is based on the ideas of Tukey (1977), though we have slightly modified his proposals. Suppose we have a sample consisting of the heights (in cm) of nine 13-year-old boys:

$$157, \ 147, \ 164, \ 152, \ 143, \ 156, \ 170, \ 159, \ 160.$$

The procedure we use is to put the observations in increasing order and then pick out (1) the minimum, or lowest, value, L; (2) the first quartile Q(1); (3) the median, or second quartile, Q(2); (4) the third quartile Q(3); (5) the maximum value, or upper bound, U.

```
500   REM  ** INPUT - DATA ARRAY **
510   REM  * VARIABLES: A$, I, I1, J, J1, N, N1, X, T$, X$ *
520   DIM X(100)

530   PRINT "GIVE TITLE AND NAME OF VARIATE"
540   INPUT "TITLE? "; T$
550   INPUT "NAME OF VARIATE? "; X$
560   INPUT "HOW MANY NUMBERS TO BE ENTERED? "; N
570   INPUT "HOW MANY AT A TIME? "; N1

580   LET I = 0
590   LET J1 = N1
600   IF N - (I+1)*N1 < 0 THEN 800
610      FOR J = 1 TO J1
620      PRINT "X("; J+I*N1; ") ";
630      INPUT X(J+I*N1)
640      PRINT
650      NEXT J
660   LET I = I + 1
670   PRINT "THE NUMBERS ENTERED WERE"
680      FOR J = 1 TO J1
690      PRINT "X("; J+(I-1)*N1; ") "; X(J+(I-1)*N1)
700      NEXT J
710   PRINT "IF CORRECT TYPE Y"
720   PRINT "OTHERWISE TYPE N"
730   INPUT A$
740   IF A$ = "Y" THEN 600
750   IF A$ <> "N" THEN 710
760   INPUT "LABEL OF NUMBER TO BE CHANGED? "; I1

770   INPUT "WHAT IS THE CORRECT NUMBER? "; X
780   LET X(I1) = X
790   GO TO 670
800   IF N-I*N1 <= 0 THEN 830
810   LET J1 = N-I*N1
820   GO TO 610
830   PRINT "ARRAY COMPLETE"

840   REM  * OUTPUT: DATA X(.), NUMBER OF DATA N *
850   REM  * TITLE T$, NAME OF VARIATE X$ *
```

```
RUN
GIVE TITLE AND NAME OF VARIATE
TITLE? BRUDDERSFORD FACTORY 1981
NAME OF VARIATE? NUMBER OF ABSENTEES IN EACH WEEK
HOW MANY NUMBERS TO BE ENTERED? 52
HOW MANY AT A TIME? 4
X(1)? 12
X(2)? 7
X(3)? 9
X(4)? 6
THE NUMBERS ENTERED WERE
X(1) 12
X(2) 7
X(3) 9
X(4) 6
IF CORRECT TYPE Y
OTHERWISE TYPE N
?N
LABEL OF NUMBER TO BE CHANGED? 2
WHAT IS THE CORRECT NUMBER? 4
THE NUMBERS ENTERED WERE
X(1) 12
X(2) 4
X(3) 9
X(4) 6
IF CORRECT TYPE Y
OTHERWISE TYPE N
?Y                                        etc.
```

Algorithm 4.1 *Input – Data Array*

These quantities have been underlined in the ranked observations:

143, 147, 152, 156, 157, 159, 160, 164, 170.

It is reasonable to take these values, which are spread out evenly through the data, as being representative of the sample. We present them in a table which also includes the sample size (or, if we had a complete population, we should give instead the total number of observations).

Sample size		9	
Median		157	
Quartiles	152		160
Extremes	143		170

Note that, with some sample sizes, the median and quartiles will fall between observations (see Section 3.7).

In producing the data for the table, we may use algorithms that we have already constructed. The assembled program is shown in Program 4.2, *Six-Number Summary*. Algorithm 4.1, *Input – Data Array*, has line numbers beginning at 500 and is copied out as it is written on page 28, but the line numbers in the other algorithms have multiples of 1000 added to them to keep them distinct. For example, *Exchange Sort* (Algorithm 3.1)

```
  10   REM   ** SIX-NUMBER SUMMARY **

 500   REM   ** INPUT - DATA SUMMARY **
 510   REM   * VARIABLES: A$, I, I1, J, J1, N, N1, X, T$, X$ *
 ...   (Algorithm 4.1)
 840   REM   * OUTPUT: DATA X(·), NUMBER OF DATA N *
 850   REM   * TITLE T$, NAME OF VARIATE X$ *

1400   REM   PUT X(·) INTO A(·) FOR SORT
1410      FOR I = 1 TO N
1420      LET A(I) = X(I)
1430      NEXT I

1500   REM   ** EXCHANGE SORT **
1510   REM   * INPUT: DATA A(·), NUMBER OF DATA N *
1520   REM   * VARIABLES: I, J, J1, W *
....   (Algorithm 3.1)
1640   REM   * OUTPUT: DATA IN INCREASING ORDER IN A(·) *

1700   REM   CALCULATE QUARTILES USING QUANTILES SUBROUTINE
1710   LET Q = 4
1720      FOR P = 1 TO 3
1730      GOSUB 3500
1740      LET Q(P) = Q0
1750      NEXT P

1800   REM   LABEL EXTREMES
1810   LET L = A(1)
1820   LET U = A(N)

2000   REM   OUTPUT SIX-NUMBER SUMMARY
2010   PRINT T$, X$
2020   PRINT
2030   PRINT "SAMPLE SIZE "; N
2040   PRINT "MEDIAN "; Q(2)
2050   PRINT "QUARTILES "; Q(1), Q(3)
2060   PRINT "EXTREMES "; L, U

3000   STOP

3500   REM   ** QUANTILES **
3510   REM   * INPUT: DATA A(·) IN INCREASING ORDER *
3520   REM   * NUMBER OF DATA N; P, Q FOR P-TH Q-TILE *
3530   REM   * VARIABLES: J1, J2, Q0 *
....   (Algorithm 3.6)
3570   REM   * OUTPUT: Q0, THE P-TH Q-TILE *
3580   RETURN
```

Program 4.2 *Six-Number Summary*

has 1000 added to its line numbers and begins at line 1500 in the program. Note that line
numbers in jump instructions in the renumbered algorithms will also need changing;
for example, the line number at the end of the conditional jump instruction on the new

line 1560 will be 1580. Relatively few instructions are needed to link the algorithms; these instructions are mostly to set appropriate input values for the algorithms. Algorithm 3.6, *Quantiles*, is used a number of times and it is convenient to use it as a subroutine. Information on the variables used in each algorithm is given in the program description. (Further discussion of combining algorithms into a program is given in Appendix A.)

Program 4.2, *Six-Number Summary*, is a complete program with input, calculation and output sections. We shall discuss the output of the six-number summary in Section 4.4.

4.4 Output of the six-number summary

We put data in a table to make clear, to ourselves or to others, the relations between numbers; the way numbers are spaced on a page brings out their interrelations. We shall consider in detail the output of the six-number summary of Section 4.3, as an example of instructing a computer to give us the output we require. The reader will quickly find (or has, very likely, done so already!) that when we output anything more complex than a simple number on a computer we must take care, or we shall find that the numbers pop up in completely unexpected parts of the screen or page.

A very real difficulty in discussing output is that we cannot rely on all computers behaving in exactly the same way. We discuss the problem assuming that the rules followed are as in *Basic BASIC* (Monro, 1978); the reader must check whether the rules for his computer are precisely the same or not.

4.4.1 The PRINT statement

For output we use the PRINT statement; it is the most versatile statement in BASIC. It takes the general form:

> *line number* PRINT *quantity delimiter . . . quantity delimiter*

The word *quantity* can represent a numerical variable, an expression, a character string or the TAB function; *delimiter* can be a comma or a semicolon and may be omitted at the end of a line. The delimiters instruct the computer to print on the same line as the last item printed, provided that there is room on the line. When there is no delimiter at the end of a line, the PRINT statement causes a jump to the next line and the cursor is positioned at the left-hand end of the line.

A line of print is divided into zones; we shall assume that there are five zones, each of 15 characters. Many output devices have available in the output line fewer characters than are used in the line on which the computer operates; a computer line may then spread over two, or even more, lines on the output device. We shall pay particular attention to a 40-character output line, as is used by many visual display units (VDUs).

4.4.2 Use of delimiters

There are two explicit delimiters available: the comma and the semicolon. A semicolon causes the next quantity printed to be immediately after, or on some machines one space after, the last quantity printed. A comma between variables produces a large space between the printed values of the variables. The first variable is printed at the left end of the first zone, and the next variable is printed at the left end of the second zone. The comma has the effect of moving the cursor to the beginning of the next zone.

The output of the *Six-Number Summary* (Program 4.2), in lines 2030–2060, has been done by what is perhaps the most obvious method: the quantities in the PRINT instructions are separated by commas. The output would be neater if we could present it in the form of the table on page 29.

The following output algorithm would produce a table of this form, provided that sufficient space is available on the output device.

```
2030    PRINT "SAMPLE SIZE",, N
2040    PRINT "MEDIAN",, Q(2)
2050    PRINT "QUARTILES", Q(1),, Q(3)
2060    PRINT "EXTREMES", L,, U
```

These PRINT statements would put the words (SAMPLE SIZE, MEDIAN, etc.) in the first zone, the lower quartile and the minimum value in the second zone, the sample size and the median in the third zone, and the upper quartile and the maximum value in the fourth zone.

If, however, the output line contains only 40 characters, a most unsatisfactory result will be obtained. The end of the third zone comes at the 48th character, which exceeds the length of the line displayed. Therefore the numbers which come in the fourth zone will be put on the next display line, and the pattern of the table will be lost. We shall consider an alternative approach.

4.4.3   Use of the TAB function

A TAB function in a PRINT statement moves the cursor to a particular column specified by the argument of the function. The form of the PRINT statement demands that the TAB function be followed by a delimiter; we recommend that a semicolon be used always. Suppose we had these two instructions:

```
10    LET X = 9
20    PRINT TAB(X); 3
```

The output process would consist of a movement of the cursor to column 9, and then the figure 3 would be printed in the next column, number 10. The argument of TAB may be an expression, such as $(I + 1)/2$; if, in a particular problem, the value of this expression when evaluated is not an integer, then the integer part of it is taken.

The desired output for the six-number summary consists of four blocks of columns. It is reasonable to allow the words to occupy the first zone of 15 characters. The numbers in the second column (i.e. the lower quartile and the minimum value) may then be printed immediately after the end of the first zone, and we can use a comma to control this. If we allow a maximum of 7 characters for these values, we need to reserve columns 16–22 for them. The next block of numbers may occupy columns 23–29, and the fourth (and final) block may occupy columns 30–37. The table will therefore fit inside 40 columns. The instructions are:

```
2030    PRINT "SAMPLE SIZE"; TAB(22); N
2040    PRINT "MEDIAN"; TAB(22); Q(2)
2050    PRINT "QUARTILES", Q(1); TAB(29); Q(3)
2060    PRINT "EXTREMES", L; TAB(29); U
```

4.4.4   Use of spaces in output

A third approach is self-explanatory. The output is identical with that of Section 4.4.3. We use '∧' to denote a space, which is obtained by pressing the space bar on the keyboard; no symbol will appear on a printer or on the VDU screen. The instructions are:

```
2030   PRINT "SAMPLE∧SIZE∧∧∧∧∧∧∧∧∧∧∧∧"; N
2040   PRINT "MEDIAN∧∧∧∧∧∧∧∧∧∧∧∧∧∧∧∧∧"; Q(2)
2050   PRINT "QUARTILES∧∧∧∧∧∧∧"; Q(1); "∧∧∧∧∧∧∧∧"; Q(3)
2060   PRINT "EXTREMES∧∧∧∧∧∧∧∧"; L; "∧∧∧∧∧∧∧∧"; U
```

Note that in all the methods we have given for printing the numbers, the machines left-justify the data (i.e. they align the most significant digits in the numbers) whereas our normal practice in writing is to round to a given number of decimal places and to right-justify the rounded numbers. On some machines the presence of a negative sign will misalign the display.

4.4.5   Rounding data

The numbers in our example of a six-number summary were simple ones, and we had no wish to round them. Very often, tables are not like this; they may, for example, contain means which have been calculated to several decimal places (the last of which may well not be accurate). Varying numbers of decimal places makes for a very untidy table layout. In these circumstances we shall want to round numbers before presenting them. It is convenient to define a function FNR to carry out this rounding; but note that when using it we must state the number of decimal places F0 that we require. The function is:

```
100    DEF FNR(X) = INT(X*(10↑F0) + 0.5)/(10↑F0)
```

Thus if we set F0 equal to 2, and X is 212.57934, then FNR(X) will equal 212.58. Setting F0 equal to -2, with the same X, would give FNR(X) equal to 200, i.e. the number would be rounded to the nearest 100. In fact this function definition is somewhat clumsy, because we have to calculate $10 \uparrow F0$ twice. If the reader is using a machine that allows him to take more than one line over a function definition, this can be avoided by giving $10 \uparrow F0$ a name before using it.

4.5   Frequency table for discrete variate with repeated values

Suppose we wish to form a frequency table from a set of observations which are the number of letters a man received each weekday over a long period:

> 1, 0, 3, 1, 2, 0, 0, 3, 1, 1, ...

A count of the frequency of each value of the variate may be obtained using a simple algorithm, *Frequency Count*, 4.3. This algorithm is for use when we have a variate taking integer values from 0 up to some upper bound. The maximum value, U, which might be obtained using *Maxmin* (Algorithm 3.5), must be stated; or, alternatively, any upper bound may be given. Algorithm 4.3 has been written assuming that an element 0 in an array is allowable, since in that case the algorithm takes a particularly simple form.

Some versions of BASIC demand that array elements begin at 1, so a note at the end of Algorithm 4.3 shows how it is modified if this is so.

```
5ØØ   REM  ** FREQUENCY COUNT **
51Ø   REM  * INPUT: DATA X(•), NUMBER OF DATA N *
52Ø   REM  * MAXIMUM VALUE OF VARIATE, U <= 5Ø *
53Ø   REM  * VARIABLES: I, W *
54Ø   DIM F(5Ø)

55Ø      FOR I = Ø TO U
56Ø      LET F(I) = Ø
57Ø      NEXT I

58Ø      FOR I = 1 TO N
59Ø      LET W = X(I)
6ØØ      LET F(W) = F(W) + 1
61Ø      NEXT I

62Ø   REM  * OUTPUT: FREQUENCY COUNTS F(•) *
```

Note: if $F(\emptyset)$ is not allowed, store $f_0, f_1, \ldots$ in $F(1)$, $F(2)$, ...Then line 59$\emptyset$ becomes

```
59Ø      LET W = X(I) + 1
```

Algorithm 4.3 *Frequency Count*

4.6   Frequency table with data grouped in classes

Data in which the values of observations are repeated are not too common. More often we need to group data into classes in order to obtain a frequency table (*ABC*, Section 1.2.2). The choice of class-intervals, so as to produce a convenient table, is not a simple decision, and we shall consider in Section 4.6.1 how best to make the choice. For the moment let us assume that the lower end-point L1, the upper end-point U1 and the class-width D1 have been chosen. The intervals are given by

$$L1 \leq x < L1 + D1, \; L1 + D1 \leq x < L1 + 2*D1, \ldots, \; L1 + (C-1)*D1 \leq x \leq U1,$$

where C is the number of classes; its calculation is described later.

```
5ØØ   REM  ** FREQUENCY TABLE **
51Ø   REM  * INPUT: DATA X(•), NUMBER OF DATA N *
52Ø   REM  * ENDPOINTS: LOWER, L1; UPPER, U1; CLASS-WIDTH D1 *
53Ø   REM  * VARIABLES: C, I, W *
54Ø   DIM F(2Ø)
55Ø   LET C = INT((U1 - L1)/D1) + 1          Number of classes
56Ø      FOR I = 1 TO C
57Ø      LET F(I) = Ø                         Zero frequencies
58Ø      NEXT I

59Ø      FOR I = 1 TO N
6ØØ      LET W = (X(I) - L1 + D1)/D1          Rank of interval in
61Ø      LET W = INT(W)                       which X(I) falls
62Ø      LET F(W) = F(W) + 1                  Frequency count
63Ø      NEXT I

64Ø   REM  * OUTPUT: FREQUENCIES F(•) *
```

Algorithm 4.4 *Frequency Table*

The frequency table is calculated in Algorithm 4.4, *Frequency Table*. The algorithm is a straightforward generalisation of Algorithm 4.3, *Frequency Count*. The rank number of the interval in which X(I) is located is determined in lines 600 and 610, and put into W. The values of W are integers, and the frequency of each value may be counted in the same way as in *Frequency Count*. The output of the results is carried out using Algorithm 4.5, *Output – Frequency Table*.

```
500  REM  ** OUTPUT - FREQUENCY TABLE **
510  REM  * INPUT: TITLE OF DATA T$, FREQUENCIES F(·) *
520  REM  * LOWER ENDPOINT L1, CLASS-WIDTH D1, NUMBER OF CLASSES C *
530  REM  * VARIABLES: I, I1, I2 *

540  PRINT T$
550  PRINT
560  PRINT "INTERVAL˄˄˄˄˄˄˄˄FREQUENCY"
570  PRINT
580      FOR I = 1 TO C - 1
590      LET I1 = L1 + D1*(I-1)
600      LET I2 = I1 + D1
610      PRINT I1; " <= X < "; I2, F(I)
620      NEXT I
630  PRINT I2; " <= X <= "; U1, F(C)
```

Algorithm 4.5 *Output – Frequency Table*

4.6.1 Choice of class-intervals

We look at two examples of data sets, in order to see what points have to be considered.

(a) *Examination marks* (%). The data might run: 61, 39, 48, 51, 22, ... with maximum 83 and minimum 22. We know without looking at the data that the possible values range from 0 to 100. A class-width of 10 is an obvious natural choice, leading us to the convenient class-intervals 0–9, 10–19, etc. Setting the lower endpoint L1 to zero, the upper endpoint U1 to 100, and the class-width D1 to 10 would produce these intervals; in the output the intervals would be described as

$$0 < = X < 10, \ 10 < = X < 20, \text{ etc.}$$

Since the maximum and minimum values in our set of data are 83 and 22, we shall obtain zero frequencies in the class-intervals 0–9, 10–19 and 90–99. There is no disadvantage in this, and the intervals suggested are probably the best choice for these data.

An automatic method of choosing class-intervals would be to set the minimum value (here 22) at the lower endpoint and the maximum value (83) at the upper end-point. If the number of classes were fixed at 12 then the class-width would be $(83–22)/12 = 5.0833333$. The intervals would be

$$22 < = X < 27.0833333, \ 27.0833333 < = X < 32.1666666, \text{ etc.}$$

This example shows at once that any automatic method, unless backed up by a substantial program, might give very awkward class-intervals. We recommend therefore that the choice of intervals should be a matter of personal judgement, using perhaps the maximum and minimum values in the data set as guides to intelligent choice.

Choice of the lower endpoint L1 and the class-width D1 fixes the sequence of class-intervals, but the computer must also be told when to stop. The *lower* endpoints of the class-intervals begin at L1 and continue, as far as possible, in the sequence L1 + D1, L1 + 2*D1, . . . , provided that they do not exceed the stated upper endpoint U1. There is no necessity for U1 to be an integral number of class-widths from L1. The final class-interval will in fact always be shorter than the others, and it is probably best to aim to make it contain no observations. In the case of the class-intervals we have recommended for the examination-marks data, the final 'interval' will consist of the single point X = 100. The final interval has been made to end at U1 in case larger values are ridiculous—as for example 103% would be in the examination marks.

(b)   *Lengths of cuckoo eggs* (mm).   The data quoted in *ABC*, Section 1.2.2, run: 22.5, 20.1, 23.3, 22.9, 23.1, . . . with maximum 25.0 and minimum 19.6. As in the manual production of a frequency table, discussed in *ABC*, an approximate class-width is given by (maximum − minimum)/10, i.e. 5.4/10 or 0.54. We thus see that a convenient choice of the class-width D1 is 0.5; and we might choose L1 to be 19.0 and U1 to be 25.5.

4.7   Two-way table from classification data

We take the opportunity here to use non-numeric data which are stored as string variables in the computer. We consider a survey in which two questions are asked, each of which has a fixed number of possible answers. (Algorithm 4.6, *Two-way Table*, allows up to five possible answers for each question, but this number is easily modified.)

To illustrate the method, suppose that we are collecting information to determine whether left-handedness is more common in women than in men. For each person in our sample we shall therefore record whether they are male (M) or female (F), and we shall ask whether they are left-handed. Of course, some people use both left and right hands for various tasks, so we must choose an unambiguous definition of 'left-handed'. For simplicity, ask only which hand they write with, and record L for left, R for right.

Our aim is to produce a 2×2 table (in this example; more generally it will be a table with a chosen number r of rows and c of columns); in the table the columns will refer to one classification factor (sex in our example) and the rows to another factor (handedness). An extra row and an extra column carry the totals:

	M	F	TOTAL
L	F(1, 1)	F(1, 2)	T1(1)
R	F(2, 1)	F(2, 2)	T1(2)
TOTAL	T2(1)	T2(2)	N

In Algorithm 4.5 we assume that the data have been stored in two string arrays, X$(·) and Y$(·). Probably the best way to input the data is to read in a line of data for each subject:

	Subject number,	Sex,	Handedness,
e.g.	3	M	R

and enter small groups of subjects at a time so that checking, by reading over the data, is made easy. Subject numbers are entered to make sure that data are not missed or entered twice, and to aid checking; it is not necessary for the subject numbers to be in order.

```
500   REM   ** TWO-WAY TABLE **
510   REM   * INPUT: DATA X$(·), Y$(·), NUMBER OF UNITS N *
520   REM   * NUMBER OF ROWS N1, OF COLUMNS N2 *
530   REM   * DESCRIPTORS: ROW R$(·), COLUMN C$(·) *
540   REM   * VARIABLES: I, J, K *
550   DIM F(5,5), R(5), C(5)

560   REM   IDENTIFY ROW AND COLUMN
570       FOR K = 1 TO N
580           FOR I = 1 TO N1
590           IF X$(K) = R$(I) THEN 620
600           NEXT I
610       GO TO 650
620           FOR J = 1 TO N2
630           IF Y$(K) = C$(J) THEN 680
640           NEXT J
650       PRINT "ERROR IN DATA ITEM "; K
660       GO TO 710

670   REM   CUMULATE FREQUENCIES
680       LET R(I) = R(I) + 1
690       LET C(J) = C(J) + 1
700       LET F(I,J) = F(I,J) + 1
710       NEXT K

720   REM   * OUTPUT: FREQUENCIES F(·,·) *
730   REM   * TOTALS: ROW R(·), COLUMN C(·) *
```

Algorithm 4.6 *Two-way Table*

In lines 560–660, the cell into which each subject should be put is identified. Then, in lines 670–710, the cell frequencies and the row and column totals are accumulated. The data should have been checked before entry into the algorithm, but if a subject cannot be allocated to a cell he is ignored and an error message is printed out.

Output of a two-way table presents difficulties if there are many rows and columns. Even with a 2×2 table (i.e. one with 2 rows and 2 columns) a simple PRINT instruction, using commas as delimiters, would give a ragged layout on a 40 character output device; four columns (two of cell frequencies, one each for totals and row descriptors) are required and these will not fit into 40 character spaces if the usual zone width of 15 characters is used. It is difficult to give a general output algorithm for the two-way table since so much depends on the length of the descriptors and the number of digits in the frequencies as well as on the numbers of rows and columns. Algorithm 4.7, *Output — 2×2 Table*, may be used for the 2×2 case. Essentially we use a TAB function to modify the zone width. Choosing a zone width M0 of 10 should give a

```
500   REM   ** OUTPUT - 2 * 2 TABLE **
510   REM   * INPUT: TITLE OF DATA T$, MARGIN MØ *
520   REM   * FREQUENCIES F(•,•), NUMBER OF UNITS N *
530   REM   * DESCRIPTORS: ROW R$(•), COLUMN C$(•) *
540   REM   * TOTALS: ROW R(•), COLUMN C(•) *
550   REM   * VARIABLES: I *

560   PRINT T$
570   PRINT
580   PRINT TAB(MØ); C$(1); TAB(2*MØ); C$(2); TAB(3*MØ); "TOTAL"
590   PRINT
600       FOR I = 1 TO 2
610       PRINT R$(I); TAB(MØ+1); F(I,1); TAB(2*MØ+1); F(I,2); TAB(3*MØ+1); R(I)
620       PRINT
630       NEXT I
640   PRINT "TOTAL"; TAB(MØ+1); C(1); TAB(2*MØ+1); C(2); TAB(3*MØ+1); N
```

Algorithm 4.7 *Output – 2 × 2 Table*

satisfactory output and allow descriptors and frequencies of up to 9 characters, although the table will have a better appearance if the descriptors are kept reasonably short (not exceeding 5 letters, say) and the frequencies are not too large. A similar approach may be used with a bigger table.

4.8   Test data for Algorithms

Algorithm 4.1, *Input-Data Array*. Input as required with $N = 6$, $N1 = 3$. Array to be 11, 21, 31, 59, 79, 99. Make mistakes on entry and correct them.

Program 4.2, *Six-Number Summary*. Input: $N = 9$, $N1 = 3$. Array to be 41, 31, 21, 17, 27, 37, 44, 33, 22. Output: $N = 9$, $Q(2) = 31$, $Q(1) = 22$, $Q(3) = 37$, $L = 17$, $U = 44$.

Algorithm 4.3, *Frequency Count*. Input: $N = 8$, $U = 6$, $X(\cdot) = (2, 3, 4, 5, 6, 5, 4, 3)$. Output: $F(\cdot) = (0, 0, 1, 2, 2, 2, 1)$.

Algorithm 4.4, *Frequency Table*. Input: $N = 12$, $L1 = 2.0$, $U1 = 10.0$, $D1 = 2.0$, $X(\cdot) = (2.3, 3.4, 4.5, 5.6, 6.7, 7.8, 8.9, 9.8, 8.7, 7.6, 6.5, 5.4)$. Output: $F(\cdot) = (2, 3, 4, 3, 0)$.

Algorithm 4.5, *Output – Frequency Table*. Output the data from Algorithm 4.4 above.

Algorithm 4.6, *Two-way Table*. Input: $N = 10$, $N1 = 2$, $N2 = 2$, $R\$(\cdot) = (L, R)$, $C\$(\cdot) = (M, F)$. Pairs of data $(X\$, Y\$): (L, M)$, (L, F), (L, M), (R, F), (L, F), (R, M), (L, F), (R, M), (L, M), (L, M). Output: $F(1, 1) = 4$, $F(1, 2) = 3$, $F(2, 1) = 2$, $F(2, 2) = 1$.

Algorithm 4.7, *Output – 2 × 2 Table*. Output the data from Algorithm 4.6 above.

4.9   Exercises

1   Run the following program which generates 100 random digits and prints out a frequency table of the digits. We would expect each digit to occur with approximately equal frequency.

```
10    REM  ** RANDOM DIGIT TEST **
20    LET N = 100
30        FOR I = 1 TO N
40        LET X(I) = INT(10*RND(1))
50        NEXT I
60    LET U = 9

500   REM  ** FREQUENCY COUNT **
...   (Algorithm 4.3)
620   REM  * OUTPUT: FREQUENCY COUNTS F(·) *

700   PRINT "VALUE", "FREQUENCY"
710       FOR I = 0 TO U
720       PRINT I, F(I)
730       NEXT I
740   END
```

2 Modify the program in question 1 so that it (1) reads N, and then reads N data into X(·); (2) calculates the maximum value U from the data. Run your program with the following data:

(a) 10, 3, 7, 2, 4   5, 6, 12, 6, 3   8, 3, 3, 4, 4   9, 2, 4, 2, 7   3, 7, 4, 2, 4   (these are the numbers of letters in the first 25 words of this chapter);

(b) the first significant digit and, separately, the third significant digit in atomic masses (which can be looked up in a table of atomic masses);

(c) the first significant digit in lists of geographical data such as lengths of rivers, heights of mountains or populations of towns.

3† [This extends the programs which were discussed in questions 1 and 2.] Write a program which:

1) inputs data and incorporates a checking procedure (use Algorithm 4.1, *Input – Data Array*);
2) calculates the maximum value U from the data (use Algorithm 3.5, *Maxmin*);
3) makes a frequency count of the data (use Algorithm 4.3, *Frequency Count*);
4) prints out the results, as in the program in question 1. The algorithms may be strung together as in a chain, or used as subroutines (see Appendix A).

Try out your program using some of the data given or described in question 2.

4 Write algorithms to carry out the following operations on a set of N data in an array X(·):

(a) given the number F0 of decimal places required, print the data in a column with all decimal points in line;

(b) given F0, print the data in a column 'right-justified', i.e. having all final digits in line.

Check using the data 16.0, 121.374, 2.9389, 21, 12.29, two decimal places being required in the output.

5† Add the algorithms *Frequency Table* (4.4) and *Output – Frequency Table* (4.5) to Program 4.2, *Six-Number Summary*, to produce a program which:

1) prints the six-number summary;

2) asks the user to input the endpoints and class-width for the frequency table;
3) prints the frequency table;
4) returns to the input of endpoints and class-width to allow these to be changed and a new table produced.

6†   Write an algorithm to produce a two-dimensional frequency table. From N pairs of data (x, y) stored in arrays $X(\cdot)$, $Y(\cdot)$ produce a two-way frequency table $F(\cdot, \cdot)$: $F(I, J)$ is to equal the number of pairs such that x falls in the Ith class of the variable X and y falls in the Jth class of the variable Y. [The algorithm is a generalisation of Algorithm 4.4, *Frequency Table*. If the reader has difficulty in doing this, he will find such an algorithm in the first part of *Output – Scattergram* (5.5).]

7   Write a flow chart for Algorithm 4.3, *Frequency Count*.

5 Inspection and summary of data using graphical methods

5.1 Introduction

The relationships between variables in a set of data can be shown up much more vividly by graphical methods than by numbers alone. But it can be very tedious to plot graphs using only pencil and paper, and often we are too lazy to do so! Our excuse is removed, however, if we have a computer to do the work for us.

Many machines now have quite elaborate facilities for graphical work. We discuss these briefly in Sections 5.7 and 5.8, but we shall first concentrate on what is available in simple BASIC. Essentially, this means using just one feature: the TAB function. Of course this has its limitations, but we think the reader will be surprised how much can be achieved using it. It is worth pointing out again the ambiguity of the TAB function. Following Monro (1978) we have adopted the convention that the effect of TAB(X) is to move the cursor to column X; therefore if the TAB function is followed by a semicolon and, say, a variable then the variable will be printed in column X + 1. This is indeed what happens on many computers; however, others when obeying the same BASIC instructions will print the variable in column X. If the reader is using a machine which works in this alternative way, a simple modification should be made in our algorithms: increase the argument of the TAB function by 1. If some rather chaotic displays are obtained on first running the algorithms, this ambiguity in TAB may well be a reason.

On a computer screen we are usually restricted to a coarse matrix of about 40×24 positions (40 columns $\times$ 24 rows); in each position we may print a character (letter, figure or other symbol). Suppose we wish to draw a scatter diagram. First we must make decisions like those required when drawing the plot on graph paper: we must fix the position and length of the axes, and we must choose appropriate scales on the axes. This is a complex set of decisions. There do exist programs for carrying out this process which ensure, for example, that scales are convenient and that plotting on the available grid produces minimum misplacement of the points. Unfortunately, one of these programs would take up more space than we can afford. So we shall adopt a simple, crude approach which, while far from perfect, will give results that are broadly satisfactory.

Within the screen area available, we choose a space, which we call the **window;** inside this the plot is framed. We fix the window width W∅ and the window height H∅, each of which is measured as a number of character spaces. For each variable we choose upper and lower limits for its values (U1 and L1 respectively for variable X, let us say); these may actually be chosen to coincide with the maximum and minimum values of the observations in a set of data. We then determine a scaling factor $S1 = W∅/(U1 - L1)$, giving the number of character spaces corresponding to each unit of the variable. Thus a point which is D units from the y-axis is $D*S1$ character spaces away from the y-axis on the screen. More precisely, it will actually be INT $(D*S1 + ∅.5)$ spaces from the y-axis, since the number of spaces must be integral. This rounding of values will lead to small misplacements of points, and straight lines may thus appear slightly distorted, although the appearance is usually acceptable. Much more precise plotting of points can be achieved using high resolution graphics (see Section 5.8).

It is useful to reserve a margin on the left-hand side of the screen in which to print numbers to indicate scale, or axis labels. We use a variable $M\emptyset$ to denote the number of character spaces allotted to the margin: 5 will usually be appropriate. This is the margin for showing information about the y-axis (in the conventional x–y representation); that for the x-axis can be printed below the plot.

5.2   Histogram

A simple histogram is easy to produce. Algorithm 5.1, *Output – Histogram*, assumes that a frequency table has already been formed from the data. The type of output is illustrated in Figure 5.1. The lower endpoint of each interval is printed on the left. Large frequencies will produce too many asterisks for a row; then the extra asterisks for an

```
500   REM   ** OUTPUT - HISTOGRAM **
510   REM   * INPUT: TITLE OF DATA T$, FREQUENCIES F(•) *
520   REM   * LOWER ENDPOINT L1, CLASS-WIDTH D1 *
530   REM   * NUMBER OF CLASSES C, ROUNDING PARAMETER FØ, MARGIN MØ *
540   REM   * VARIABLES: I, J, W *

550   DEF FNR(X) = INT(X*(10↑FØ) + 0.5)/(10↑FØ)
560   PRINT T$
570       FOR I = 1 TO C
580       LET W = L1 + (I-1)*D1                    Calculate and print
590       LET W = FNR(W)                           lower class interval
600       PRINT W; "-"; TAB(MØ); "I";
610       IF F(I) = 0 THEN 650
620           FOR J = 1 TO F(I)                    Print bar of
630           PRINT "*";                           histogram
640           NEXT J
650       PRINT
660       NEXT I
```

Algorithm 5.1 *Output – Histogram*

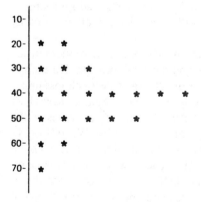

Figure 5.1

interval affected in this way will be printed on the next line and this spoils the histogram. We suggest two alternative approaches for this case: (a) scale all the frequencies so that the maximum will fit in the space available, or (b) test whether each frequency exceeds a number which is a few units less than the maximum space available; if it does, print '+N' at the end of the line of asterisks, N being the number of asterisks that there is not room to print.

5.3 Stem and leaf diagram

Tukey (1977) has proposed a form of diagram which represents the distribution of a set of numbers in a similar way to a histogram, yet also retains most, if not all, of the information on individual values. We may think of it as a concise way of writing down a set of numbers. As an example, a stem-and-leaf diagram for the (ordered) numbers 21, 23, 25, 32, 32, 34, 35, 36, 38, 39, 40, 43, 44, 45, 51, 55, 67 is:

```
2 | 135
3 | 2245689
4 | 0345
5 | 15
6 | 7
```

Thus 21 is represented on the first row, or **stem** as Tukey calls it, by the stem label '20', which we replace by '2', together with '1' which is a **leaf** on the stem. The other two

```
500   REM   ** OUTPUT - STEM AND LEAF **
510   REM   * INPUT: DATA A(·) IN ASCENDING ORDER *
520   REM   * DATA INTEGERS, NOT MORE THAN 3 DIGITS *
530   REM   * LOWER ENDPOINT L1, CLASS-WIDTH D1 *
540   REM   * NUMBER OF DATA N *
550   REM   * VARIABLES: I, I1, I2, W, W1, W2, A$, L$ *

560   LET A$ = ""                          Leaf string zeroed
570   LET I1 = L1                          Interval endpoints
580   LET I2 = L1 + D1

590      FOR I = 1 TO N
600      LET W = A(I)
610      LET W1 = 10*INT(W/10)             Find stem label
620      LET L$ = STR$(W - W1)             Find leaf
630      IF W < I2 THEN 700                Completion of leaf?
640      LET W2 = INT(I1/10)               Stem label
650      PRINT W2; TAB(3); "*"; A$         Print label and leaves
660      LET I1 = I2                       Reset interval limits
670      LET I2 = I1 + D1
680      LET A$ = ""
690      IF W1 > I1 THEN 640               Interval empty
700      LET A$ = A$ + L$                  Add leaf to stem
710      NEXT I

720   LET I1 = INT(I1/10)
730   PRINT I1; TAB(3); "*"; A$            Print last stem
```

Algorithm 5.2 *Output – Stem and Leaf*

leaves, 3 and 5, on this stem correspond to the numbers 23 and 25. The next stem has the label '30', represented by the '3' on the left, and the seven leaves on this stem correspond to the seven numbers of the set of data which are in the thirties. The idea is discussed at length, and elaborated, by Tukey (1977).

For input to Algorithm 5.2, *Output – Stem and Leaf*, the data are assumed to be in ascending order, having been sorted; each observation is assumed to be an integer of two or three digits. The diagram divides the data into classes. It is necessary to input the desired lower endpoint and class-width; in the above example the lower endpoint was 20 and the class-width 10. A value of 10 is particularly easy to interpret but, depending on the data, Tukey also recommends 2, 5, 20, 50 as possible class-widths. The algorithm can be used with a wide range of class-width values, and the reader may like to experiment with this on his own machine.

The algorithm takes the ith element, finds its stem label (line 610), and then finds its leaf which is stored as a string variable (line 620). The leaves for a particular stem are assembled (line 700) using the operation of adding string variables; the complete set of leaves for a particular stem is printed out in one operation (line 650).

Data are most easily interpreted when they come from a symmetrical distribution. One use of a stem and leaf diagram is to check whether a distribution is symmetrical; if it is not, the statistician considers transforming the original data in order to obtain a more symmetrical distribution. In Section 5.4 we introduce an algorithm (5.3) for making transformations. Also, Algorithm 5.2 is rather restrictive in the type of data it will accept: it requires integers of two or three digits, and data may therefore need to be transformed into a form acceptable for this algorithm.

5.4   Transformations

Algorithm 5.3, *Transformations*, is for interactive use. It is illustrative only, for by no means all the possible useful transformations have been included. It has been written for use with Algorithm 5.2, *Output – Stem and Leaf*, the results of all the transformations being expressed as integers.

5.5   Box and whisker plot

Tukey (1977) proposed another form of diagram, the box and whisker plot, which may be regarded as a diagrammatic presentation of the six-number summary (see Section 4.3). (It would strictly be more accurate to say five-number summary, since the sample size is not shown unless incorporated as a separate line of output.)

An example is shown in Figure 5.2. Basically it is a representation of the five numbers on a line, but the portion between the first quartile (32) and the third quartile (61), which

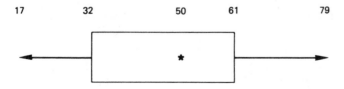

Figure 5.2

```
500   REM  ** TRANSFORMATIONS **
510   REM  * INPUT: DATA X(•), NUMBER OF DATA N *
520   REM  * VARIABLES: A, B, I, W *
530   DIM U(100)

540   PRINT "0. NO TRANSFORMATION"
550   PRINT "1. LINEAR TRANSFORMATION"
560   PRINT "2. LOG TRANSFORMATION"
570   PRINT "3. SQUARE ROOT TRANSFORMATION"
580   PRINT "4. REPLACE BY ORIGINAL DATA"
590   INPUT "ENTER TRANSFORMATION CODE "; W

600   IF W = 0 THEN 940
610   IF W <> 1 THEN 710
620   PRINT "TRANSFORMATION NEW X = INT(A+B*X)"
630   INPUT "WHAT IS A? "; A
640   INPUT "WHAT IS B? "; B
650      FOR I = 1 TO N
660      LET W = X(I)
670      LET U(I) = W
680      LET X(I) = INT(A+B*W)
690      NEXT I
700   GO TO 940

710   IF W <> 2 THEN 790
720   PRINT "TRANSFORMATION NEW X = INT(100*LOG(X+1))"
730      FOR I = 1 TO N
740      LET W = X(I)
750      LET U(I) = W
760      LET X(I) = INT(100*LOG(W+1))
770      NEXT I
780   GO TO 940

790   IF W <> 3 THEN 870
800   PRINT "TRANSFORMATION NEW X = INT(SQR(X))"
810      FOR I = 1 TO N
820      LET W = X(I)
830      LET U(I) = W
840      LET X(I) = INT(SQR(W))
850      NEXT I
860   GO TO 940

870   IF W <> 4 THEN 920
880      FOR I = 1 TO N
890      LET X(I) = U(I)
900      NEXT I
910   GO TO 940

920   PRINT "CODE INVALID; RE-ENTER"
930   GO TO 540

940   REM  * OUTPUT: TRANSFORMED DATA IN X(•) *
950   REM  * ORIGINAL DATA IN U(•) *
```

Algorithm 5.3 *Transformations*

contains half the population, is thickened and represented as a box. Whiskers reach out from the left and right sides of the box to the extremes (17 and 79). The median (50) is denoted by an asterisk within the box. We give an algorithm *Output – Box and Whisker Plot*, 5.4.

```
500    REM  ** OUTPUT - BOX AND WHISKER PLOT **
510    REM  * INPUT: SIX-NUMBER SUMMARY, N, L, Q(1), Q(2), Q(3), U *
520    REM  * WINDOW WIDTH WØ; ENDPOINTS L1, U1 *
530    REM  * ROUNDING PARAMETER FØ *
540    REM  * VARIABLES: I, S1 *
550    DIM P(5)

560    DEF FNP(X) = INT(S1*(X - L1)) + 1          Distance function
570    DEF FNR(X) = INT(X*(10↑FØ) + Ø.5)/(10↑FØ)  Rounding function
580    LET Q(Ø) = L
590    LET Q(4) = U
600    LET S1 = WØ/(U1 - L1)                       Scaling factor

610    PRINT "N = "; N
620    PRINT
630        FOR I = Ø TO 4
640        LET P(I) = FNP(Q(I))                    Position of quantiles
650        LET Q(I) = FNR(Q(I))                    Round quantiles
660        PRINT TAB(P(I) - 1); Q(I);              Print quantile values
670        NEXT I
680    PRINT
690    PRINT
700    PRINT TAB(P(1));                            Draw top of box
710        FOR I = P(1) TO P(3)
720        PRINT "-";
730        NEXT I
740    PRINT
750    PRINT TAB(P(1)); "I"; TAB(P(3)); "I"        Side of box
760    PRINT TAB(P(Ø)); "(";                       Draw left whisker
770        FOR I = P(Ø) + 1 TO P(1)
780        PRINT "-";
790        NEXT I
800    PRINT TAB(P(2)); "*"; TAB(P(3));            Print median mark
810        FOR I = P(3) TO P(4) - 1                Draw right whisker
820        PRINT "-";
830        NEXT I
840    PRINT ")"
850    PRINT TAB(P(1)); "I"; TAB(P(3)); "I"        Side of box
860    PRINT TAB(P(1));                            Draw bottom of box
870        FOR I = P(1) TO P(3)
880        PRINT "-";
890        NEXT I
```

Algorithm 5.4 *Output – Box and Whisker Plot*

To scale the plot, we must input the window width WØ (given as the number of character spaces) of the output device, and the upper and lower endpoints, U1 and L1, for the data. These last two values may be set equal to the maximum (U) and minimum

(L) values of the data if the plot is to occupy the whole window width. Endpoints distinct from the extreme data values are useful if box and whisker plots for a number of populations are to be compared; the same endpoints may be used in plotting each population. A rounding parameter F0, which is the number of decimal places in the rounded values, must also be input; unless they are rounded, the values in the six-number summary are likely to run into each other when output.

Algorithm 5.4 begins by finding the scaling factor S1. Then the positions P(I) of the five numbers are found, measured as distances from the lower endpoint, by using the function FNP(X) defined in line 560. A number of statements can be written more concisely if we call the minimum value (L) in the data Q(0) and the maximum (U) in the data Q(4); all five summary numbers will then be of the form Q(I), I = 0, 1, 2, 3, 4. Rounding the five numbers, using the rounding function FNR(X), is done before the five numbers are printed in line 660.

5.6 Scatter diagram

Scatter diagrams should always be used when investigating the relation between two variables. We illustrate in Figure 5.3 the type of output that is obtained from Algorithm 5.5, *Output – Scattergram*.

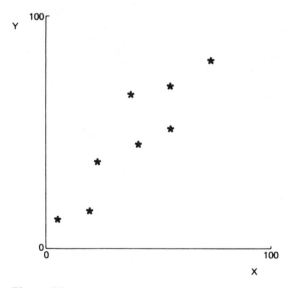

Figure 5.3

The method adopted is, first, to classify the data into a two-way frequency table, and then to print a mark on the plot corresponding to cells which have a non-zero frequency. When the frequency is unity an asterisk is printed, and when it is greater than unity the actual value of the frequency is printed.

Window width W0 and height H0 must be input. Most screens allow 24 (height) × 40 (width) print positions; thus to allow space for text outside the window, and to guard against errors, it is wise not to allow H0 to exceed 20 nor W0 to exceed 30. The margin

```
500   REM   ** OUTPUT - SCATTERGRAM **
510   REM   * INPUT: DATA X(•), Y(•), NUMBER OF UNITS N *
520   REM   * LOWER ENDPOINTS L1, L2; UPPER ENDPOINTS U1, U2 *
530   REM   * WINDOW WIDTH WØ, HEIGHT HØ, MARGIN MØ *
540   REM   * VARIABLES: D1, D2, H1, I, J, S1, S2, W1, W2, W3 *
550   DIM F(2Ø,2Ø)

560   LET S1 = WØ/(U1 - L1)                          Scaling factor
570   LET S2 = HØ/(U2 - L2)                          Scaling factor

580   REM   FORM FREQUENCY TABLE
590       FOR I = 1 TO WØ                            Zeroes frequency
600           FOR J = 1 TO HØ                        matrix
610           LET F(I,J) = Ø
620           NEXT J
630       NEXT I

640   LET D1 = L1 - 1/S1
650   LET D2 = L2 - 1/S2
660       FOR I = 1 TO N
670       LET W1 = (X(I) - D1)*S1
680       LET W2 = (Y(I) - D2)*S2
690       LET W1 = INT(W1 + Ø.5)
700       LET W2 = INT(W2 + Ø.5)
710       LET F(W1,W2) = F(W1,W2) + 1
720       NEXT I

730   REM   PLOT DATA
740   PRINT U2;                                      Maximum value of y
750   LET H1 = HØ + 1
760       FOR J = 1 TO HØ
770       IF J < HØ THEN 79Ø
780       PRINT L2;                                  Minimum value of y
790       PRINT TAB(MØ); "I";
800           FOR I = 1 TO WØ
810           LET W3 = F(I,H1-J)
820           IF W3 <= Ø THEN 87Ø
830           IF W3 > 1 THEN 86Ø
840           PRINT TAB(MØ+I); "*";
850           GO TO 87Ø
860           PRINT TAB(MØ+I); W3;
870           NEXT I
880       PRINT
890       NEXT J

900   PRINT TAB(MØ); "-";
910       FOR I = 1 TO WØ
920       PRINT "-";
930       NEXT I
940   PRINT
950   PRINT TAB(MØ+1); L1; TAB(MØ+WØ); U1
```

Algorithm 5.5 *Output – Scattergram*

size, M0, must also be input, and a convenient value for it is 5. Apart from the actual data, the other input requirements are the upper and lower endpoints for each variable.

The method used in the first part of the algorithm (lines 580–720) is a simple extension to two variables of the Algorithm 4.4, *Frequency Table*. The second part of Algorithm 5.5 prints the axes, the numerical values at the extremes of the axes, and the points.

5.7 Scatter diagram using cursor controls

Many computers offer special facilities for graph plotting that are additional to simple BASIC. We shall now make use of instructions that control the movement of the cursor in a more flexible way than does the TAB function. We regard the computer screen as a sheet of graph paper and by means of the cursor controls we move the cursor to any point we wish and print a symbol there. Algorithm 5.6, *Output – Scattergram (Graphics)*, includes statements such as 580 PRINT 'clear screen', which must be translated into statements that are acceptable to the reader's particular machine and have the stated effect when used on it. The reader will have to consult the machine manual to find these statements. When writing out this algorithm, we have allowed ourselves to put more than one instruction on a line, to save space. This is a facility that is available in many versions of BASIC.

```
500  REM  ** OUTPUT - SCATTERGRAM (GRAPHICS) **
510  REM  * INPUT: DATA PAIRS X(·), Y(·); NUMBER OF UNITS N *
520  REM  * LOWER LIMITS L1, L2; UPPER LIMITS U1, U2 *
530  REM  * WINDOW WIDTH W0, HEIGHT H0, MARGIN M0 *
540  REM  * VARIABLES: I, I1, I2, K, S1, S2, W *

550  REM  CALCULATE SCALE FACTORS
560  S1 = W0/(U1 - L1) : S2 = H0/(U2 - L2) : W = M0 - 1
570  REM  DRAW AXES AND PRINT LIMITS
580  PRINT "clear screen" : PRINT U2; TAB(W); "| ";      Upper limit of y
590  FOR I = 1 TO H0 : PRINT "cursor down | ";: NEXT I    Y-axis
600  PRINT : PRINT L2; TAB(W); "L";                       Lower limit of y
610  FOR I = 1 TO W0 : PRINT " ";: NEXT I : W = W - 1 : PRINT   X-axis
620  PRINT : PRINT TAB(W); L1; TAB(W + W0); U1            Limits of x

630  FOR K = 1 TO N
640  REM  CALCULATE CURSOR POSITION
650  I1 = (X(K) - L1)*S1 : I1 = INT(I1 + 0.5) + W
660  I2 = (U2 - Y(K))*S2 : I2 = INT(I2 + 0.5)
670  REM  POSITION CURSOR AND PRINT
680  PRINT "cursor home"                                  Place cursor at top left
690  FOR I = 0 TO I1 : PRINT "cursor right";: NEXT I      X-movement
700  FOR I = 0 TO I2 : PRINT "cursor down";: NEXT I       Y-movement
710  PRINT "*" : NEXT K                                   Print point
720  PRINT "cursor home"
730  FOR I = 1 TO H0 + 4 : PRINT "cursor down";: NEXT I
```

Algorithm 5.6 *Output – Scattergram (Graphics)*

Input is identical with that of Algorithm 5.5. The algorithm first calculates the scaling factors, and then the axes and associated numerical values are printed by lines 570–620. The coordinate values of each point are scaled, and the position on the plot scaled, in lines 630–660. Instruction 680 places the cursor at the top left-hand corner of the screen, and lines 690–700 move it to its correct position where a symbol is printed (line 710). (Note that the 'origin' for the cursor is at the *top* left-hand corner of the screen, whereas on a sheet of graph paper we should usually take the origin near the bottom left-hand corner.) When all the points have been printed the cursor is moved to a point below the diagram, so that any further output does not overwrite the completed scattergram.

5.8   High resolution graphics

We have seen that it is generally possible to position the cursor at any of 40 horizontal positions, and any of 24 vertical positions; the printing unit at a 'position' is 7×8 dots. Some machines allow the programmer to position the cursor on any one of these actual dots, so giving a range of 280×192 printing locations. This enables the programmer to produce much finer pictures and to achieve much closer approximations to continuous straight lines. In conjunction with this high resolution, there are inbuilt functions which will draw a straight line between two points whose coordinates are given. This opens up enormous possibilities in representing data pictorially. Conventional histograms consisting of rectangles are easy to construct, and pie charts also become possible; even on moderately priced machines the segments of pie charts can be exhibited in contrasting colours.

5.9   Test data for Algorithms

Algorithm 5.1, *Output – Histogram*. Input: T\$ = 'A   TITLE',   C = 5,   F(·) = (2, 4, 6, 12, 6), L1 = 5.14, D1 = 2.5, F0 = ·1, M0 = 10.
Algorithm 5.2, *Output – Stem and Leaf*. Input: N(= 30) numbers of the form A(I) = A1 + A2*I, where A1 = 1, A2 = 3 and I = 1 to N. L1 = 0, D1 = 10.
Algorithm 5.3, *Transformations*. Input: N = 5, X(·) = (9, 16, 100, 200, 500). Code = 1; A = 7,   B = 0.1;   Output: X(·) = (7,   8,   17,   27,   57).   Code = 2;   Output: X(·) = (230, 283, 461, 530, 621). Code = 3; Output: X(·) = (3, 4, 10, 14, 22). Code = 4; Output: X(·) = (9, 16, 100, 200, 500), provided that one of the earlier codes has been used. Code = 5; Output: error message.
Algorithm 5.4, *Output – Box and Whisker Plot*. Input: N = 100, L = 20, Q(1) = 40, Q(2) = 51.5, Q(3) = 73, U = 120, L1 = 10, U1 = 150, F0 = 1, W0 = 40.
Algorithm 5.5, *Output–Scattergram*. Input: N = 6, L1 = 1, L2 = 1, U1 = 15, U2 = 20, W0 = 30, H0 = 20, M0 = 4. Pairs (x, y) = (2, 2), (7, 2), (7, 7), (12, 2), (12, 7), (12, 12). Output: triangle of points. (Change dimension statement 550 to F(30, 20).)

5.10   Exercises

1   At the beginning of Algorithm 5.2, *Output – Stem and Leaf*, introduce a jump to a subroutine which:
  i) checks that the data in A(·) are in ascending order, and prints a message if the check fails;

ii) checks that the lower endpoint L1 is less than the minimum of the data, i.e. A(1), and prints a message if the check fails.

2† Write a program which:
i) inputs N data, each of which is an integer not exceeding three digits, into an array X(·);
ii) sorts the data and puts them in A(·) in ascending order of size;
iii) finds the maximum (U) and minimum (L) of the data;
iv) uses *Output – Stem and Leaf* (Algorithm 5.2) with lower endpoint L1 = 2 and class width D1 = INT((U − L)/12).

3† Write a program which:
i) inputs N data into an array X(·);
ii) calculates the six-number summary (see Section 4.3);
iii) prints a box and whisker plot (see Section 5.5);
iv) allows a further set of data to be entered, and an additional box and whisker plot to be printed, which may be compared with the previous plot.

4 Use *Output – Scattergram* (Algorithm 5.5), to plot a *time-series*, i.e. to show how the record y varies when x represents equally-spaced time intervals, as in the following data on the quarterly production of a certain animal feeding stuff (in tonnes).

Year/Quarter	Production	Year/Quarter	Production
1977/1	32.5	1979/1	35.4
2	42.2	2	48.0
3	35.1	3	42.6
4	41.0	4	49.8
1978/1	34.7	1980/1	38.5
2	45.5	2	52.3
3	37.6	3	47.3
4	44.2	4	54.3

5 Simulate how the proportion of heads varies as a coin is tossed repeatedly, the probability of obtaining a head at each single toss being P.

1) In the following algorithm, the number of heads after N tosses is given by the final value of T:

```
100    LET T = 0
110        FOR I = 1 TO N
120        LET U0 = RND(1)
130        IF U0 > P THEN 150
140        LET T = T + 1
150        NEXT I
```

2) Using a window width of (say) 40 plot 'X' at INT(40*P + 0.5), to mark the expected proportion of heads. Repeat this at intervals of 10 rows. Plot '*' at INT(40*T + 0.5) after each odd-numbered toss (i.e. I = 1, 3, 5, . . .). [This plot may be done either by

using the TAB function or by printing the appropriate number of spaces, using " ",
before printing the asterisk.]

3) Run the program with P = 0.5, and check whether the proportion of heads stabilises
at 0.5 as N becomes large.

6   Run Algorithm 5.2, *Output – Stem and Leaf,* using as input data *either* (a) the atomic
masses of the elements taken from a table in a chemistry book (take the stem in units of
10) *or* (b) random numbers between 0 and 100 generated using the RND function, as
described in Section 7.3, page 67, except that the necessary instruction will be:
```
510   LET W = INT(100*W)
```

7   Use appropriate algorithms to construct a scatter diagram for the following data,
which give the mathematics (x) and physics (y) marks of eleven students A–L in the
same class:

	A	B	C	D	E	F	G	H	J	K	L
x:	41	37	38	39	49	47	42	34	36	48	29
y:	36	20	31	24	37	35	42	26	27	29	23

8†   Repeat question 7 for suitable pairs of data from Appendix C, e.g. (a) H and W,
(b) R and C, separately for the two sexes.

6 Computation of variance and correlation coefficient

6.1 Variance by methods based on standard formulae

Texts on statistics usually give two alternative and equivalent formulae for calculating the estimated population variance, s^2, based on a sample of N observations:

$$s^2 = \sum_{i=1}^{N} (x_i - \bar{x})^2/(N-1); \tag{1}$$

$$s^2 = \left(\sum_{i=1}^{N} x_i^2 - (\sum x_i)^2/N \right)/(N-1). \tag{2}$$

Note that we are using a divisor $(N-1)$. This is appropriate when estimating the variance of a population, given a sample of N observations from the population (ABC, Section 10.5), and is what we usually require. If the N observations do actually constitute a complete population, the variance is given by similar formulae in which N replaces $(N-1)$. Formula (1) is the direct definition of variance, while in (2) the sum of squares $\sum (x_i - \bar{x})^2$ has been expanded and rearranged in a form that is often more convenient for desk calculation. The term $(\sum x_i)^2/N$ is called a **correction term** and the numerator $\sum x_i^2 - (\sum x_i)^2/N$ is called a **corrected sum of squares.**

```
500   REM   ** MEAN AND VARIANCE (DEVIATION METHOD) **
510   REM   * INPUT: DATA X(•), NUMBER OF DATA N *
520   REM   * VARIABLES: I, M, T, V, W *

530   LET T = Ø
540       FOR I = 1 TO N
550       LET T = T + X(I)
560       NEXT I
570   LET M = T/N

580   LET T = Ø
590       FOR I = 1 TO N
600       LET W = X(I) - M
610       LET T = T + W*W
620       NEXT I
630   LET V = T/(N - 1)

640   REM   * OUTPUT: MEAN M, ESTIMATED VARIANCE V *
```

Algorithm 6.1 *Mean and Variance (Deviation Method)*

Algorithm 6.1, *Mean and Variance (Deviation Method)*, is constructed by the method of formula (1). This algorithm is straightforward and easy to understand. It might be criticised on grounds of efficiency, as it requires two passes through the data: one to calculate the mean and a second to calculate the squared deviations from the mean, which then allow us to calculate the variance.

Formula (2) looks as if it might be more efficient since the calculation requires only a single pass through the data. Unfortunately it has a serious disadvantage: it can give inaccurate results. The subtraction of the correction term from the uncorrected sum of squares $\sum x_i^2$ may lead to a substantial loss of significant figures because it takes the difference of two numbers, both of which may be large. We illustrate the nature of the inaccuracy by an example.

Suppose we wish to estimate the variance from a sample consisting of the numbers 1000, 1003, 1006, 1007, 1009, using a computer which expresses floating point numbers correct to 6 decimal digits. In Table 6.1 we follow through the calculation of the variance using formula (2), assuming (a) that the numbers are truncated, or (b) that they are rounded, in order to fit into the space available for each number. We give also the exact calculation; and we see that the result is disconcerting. The errors in the results (7.5 or 15, instead of the correct value of 12.5) are substantial; and other sets of numbers could be chosen that would give even more dramatic errors. An operator using a desk calculator would immediately 'code' the data by subtracting 1000 from each observation before calculating a variance, and this would avoid the error. But one cannot rely on data being first scrutinised like this when a computer is used. We must conclude that the method based on formula (2) can be unreliable and so should not be used.

Table 6.1

	Exact	Truncated	Rounded
x_1^2	1000000	100000 E1	100000 E1
x_2^2	1006009	100600 E1	100601 E1
x_3^2	1012036	101203 E1	101204 E1
x_4^2	1014049	101404 E1	101405 E1
x_5^2	1018081	101808 E1	101808 E1
$A = \sum x^2$	5050175	505015 E1	505018 E1
$(\sum x)^2$	25250625	252506 E2	252506 E2
$B = (\sum x)^2/5$	5050125	505012 E1	505012 E1
$A - B$	50	3 E1 = 30	6 E1 = 60
Variance, $(A - B)/4$	12.5	7.5	15

6.2   An updating procedure for calculating mean and variance

The method we recommend for calculating a variance (and the corresponding mean) avoids the disadvantages of both the methods based on standard formulae given in Section 6.1. It is accurate and, since it only requires one pass through the data, it is fast. This method updates the value of the mean and the sum of squares of deviations as each observation is introduced into the calculation. However, it is a less obvious and straightforward method than that of Algorithm 6.1, based on the direct definition of variance; some readers may therefore prefer that earlier method on the grounds that it is easier to understand.

Write m_i for the mean of the first i observations $(= \sum_{r=1}^{i} x_r/i)$, and s_i for the sum of squares of deviations of the first i observations about their mean $(= \sum_{r=1}^{i} (x_r - m_i)^2)$. We shall prove, in Section 6.3, the recurrence relations

$$m_i = [(i-1)m_{i-1} + x_i]/i;$$
$$s_i = s_{i-1} + (i-1)(x_i - m_{i-1})^2/i.$$

Algorithm 6.2, *Mean and Variance (Update Method)*, calculates the mean and variance of a set of data, based on these recurrence relations. Note that the deviation W is calculated (in line 560) before the mean M is updated. The program statements keep very close to the formulae above for the sake of clarity. It would be more efficient computing to avoid repeated calculations by replacing $I - 1$ and $X(I)$ by variables I1 and W1, respectively, at the beginning of the loop. This is done in the similar but longer Algorithm 6.4.

```
500  REM  ** MEAN AND VARIANCE (UPDATE METHOD) **
510  REM  * INPUT: DATA X(•), NUMBER OF DATA N *
520  REM  * VARIABLES: I, M, T, V, W *

530  LET M = 0
540  LET T = 0
550      FOR I = 1 TO N
560      LET W = X(I) - M
570      LET M = ((I-1)*M + X(I))/I
580      LET T = T + W*W*(I-1)/I
590      NEXT I
600  LET V = T/(N - 1)

610  REM  * OUTPUT: MEAN M, ESTIMATED VARIANCE V *
```

Algorithm 6.2 *Mean and Variance (Update Method)*

6.3* Proof of recurrence relations for mean and variance

We require to prove the two relations given in Section 6.2. The proof of that for m_i is straightforward. If we multiply both sides of the equation by i, we have

$$im_i = (i-1)m_{i-1} + x_i.$$

Each side is now the sum of the i observations, since

$$im_i = \sum_{r=1}^{i} x_r \quad \text{and} \quad (i-1)m_{i-1} = \sum_{r=1}^{i-1} x_r.$$

Before proving the relation for s_i, note that

$$i(m_i - m_{i-1}) = im_i - im_{i-1}$$
$$= (i-1)m_{i-1} + x_i - im_{i-1} \quad \text{(from the result for means)}$$
$$= x_i - m_{i-1}.$$

Now

$$s_{i-1} = \sum_{r=1}^{i-1} (x_r - m_{i-1})^2 = \sum_{r=1}^{i} (x_r - m_{i-1})^2 - (x_i - m_{i-1})^2$$

$$= \sum_{r=1}^{i} [(x_r - m_i) + (m_i - m_{i-1})]^2 - (x_i - m_{i-1})^2$$

$$= \sum_{r=1}^{i} (x_r - m_i)^2 + \sum_{r=1}^{i} (m_i - m_{i-1})^2 - (x_i - m_{i-1})^2.$$

Here the cross-product term disappears since

$$\sum_{r=1}^{i} (x_r - m_i)(m_i - m_{i-1}) = (m_i - m_{i-1}) \sum_{r=1}^{i} (x_r - m_i)$$

$$= (m_i - m_{i-1}) \times 0 = 0.$$

Thus

$$s_{i-1} = s_i + \sum_{r=1}^{i} (x_i - m_{i-1})^2 / i^2 - (x_i - m_{i-1})^2$$

$$= s_i + (x_i - m_{i-1})^2 / i - (x_i - m_{i-1})^2$$

$$= s_i - (i-1)(x_i - m_{i-1})^2 / i$$

which leads at once to the result in Section 6.2.

6.4   Calculation of sums of squares and products of two variates

To calculate the product-moment correlation coefficient of two variates X and Y, we need the corrected sums of squares, $\sum (x_i - \bar{x})^2$ and $\sum (y_i - \bar{y})^2$, and the corrected sum of products $\sum (x_i - \bar{x})(y_i - \bar{y})$. Algorithms 6.1 and 6.2 include the calculation of a corrected sum of squares; Algorithm 6.1 extends in an obvious way to give Algorithm 6.3, *SS and SP (Deviation Method)*, which includes the calculation of the corrected sum of products. The results are stored in a sum-of-squares-and-products matrix $S(\cdot, \cdot)$:

$$\begin{bmatrix} \sum (x_i - \bar{x})^2 & \sum (x_i - \bar{x})(y_i - \bar{y}) \\ \sum (x_i - \bar{x})(y_i - \bar{y}) & \sum (y_i - \bar{y})^2 \end{bmatrix}.$$

Thus $S(1, 1) = \sum(x_i - \bar{x})^2$, $S(2, 2) = \sum(y_i - \bar{y})^2$ and $S(1, 2) = \sum(x_i - \bar{x})(y_i - \bar{y}) = S(2, 1)$. Note that Algorithms 6.3 and 6.4 store both off-diagonal elements although only one need be stored since they are equal. The reader may remove, if he wishes, the single line at the end of each algorithm which stores the second off-diagonal element.

For large symmetric matrices it is of course sensible to work with only the lower (or only the upper) triangle of elements in order to economise on storage space.

```
500   REM   ** SS AND SP (DEVIATION METHOD) **
510   REM   * INPUT: DATA X(•), Y(•); NUMBER OF UNITS N *
520   REM   * VARIABLES: I, M1, M2, T1, T2, T3, W1, W2 *
530   DIM S(2,2)

540   REM   CALCULATION OF MEANS
550   LET T1 = 0
560   LET T2 = 0
570       FOR I = 1 TO N
580       LET T1 = T1 + X(I)
590       LET T2 = T2 + Y(I)
600       NEXT I
610   LET M1 = T1/N
620   LET M2 = T2/N

630   REM   CALCULATION OF SS AND SP
640   LET T1 = 0
650   LET T2 = 0
660   LET T3 = 0
670       FOR I = 1 TO N
680       LET W1 = X(I) - M1
690       LET W2 = Y(I) - M2
700       LET T1 = T1 + W1*W1          Sum (x − x̄)²
710       LET T2 = T2 + W2*W2          Sum (y − ȳ)²
720       LET T3 = T3 + W1*W2          Sum (x − x̄)(y − ȳ)
730       NEXT I
740   LET S(1,1) = T1
750   LET S(2,2) = T2
760   LET S(1,2) = T3
770   LET S(2,1) = T3                  If required

780   REM   * OUTPUT: MEANS M1, M2; SS AND SP MATRIX S(•,•) *
```

Algorithm 6.3 *SS and SP (Deviation Method)*

To extend the updating method of Section 6.2, we need to find a recurrence relation for the sum of products. By analogy with the definition and relation for s_i, we define the sum of products of i pairs of observations by

$$p_i = \sum_{r=1}^{i} (x_r - m_{1,i})(y_r - m_{2,i}),$$

and it can be shown that

$$p_i = p_{i-1} + (i-1)(x_i - m_{1,i-1})(y_i - m_{2,i-1})/i.$$

The result is used in Algorithm 6.4, *SS and SP (Update Method)* overleaf.

6.5 Input of bivariate data

Bivariate data are pairs of measurements, each pair being recorded on a single unit of a population or sample. Thus we might measure heights and weights of a sample of men;

```
500   REM   ** SS AND SP (UPDATE METHOD) **
510   REM   * INPUT: DATA X(·), Y(·); NUMBER OF UNITS N *
520   REM   * VARIABLES: I, J, M1, M2, T1, T2, T3, W1, W2, W3, W4 *
530   DIM S(2,2)

540   LET M1 = Ø
550   LET M2 = Ø
560   LET T1 = Ø
570   LET T2 = Ø
580   LET T3 = Ø
590      FOR I = 1 TO N
600      LET J = I - 1
610      LET W1 = X(I)
620      LET W2 = Y(I)
630      LET W3 = W1 - M1
640      LET W4 = W2 - M2
650      LET M1 = (J*M1 + W1)/I
660      LET M2 = (J*M2 + W2)/I
670      LET T1 = T1 + J*W3*W3/I
680      LET T2 = T2 + J*W4*W4/I
690      LET T3 = T3 + J*W3*W4/I
700      NEXT I
710   LET S(1,1) = T1
720   LET S(2,2) = T2
730   LET S(1,2) = T3
740   LET S(2,1) = T3

750   REM   * OUTPUT: MEANS M1, M2; SS AND SP MATRIX S(·,·) *
```

Lines 670–690 annotations:
$$\text{Sum } (x - \bar{x})^2$$
$$\text{Sum } (y - \bar{y})^2$$
$$\text{Sum } (x - \bar{x})(y - \bar{y})$$

Line 740 annotation: If required

Algorithm 6.4 *SS and SP (Update Method)*

```
500   REM   ** INPUT - BIVARIATE DATA (SMALL SET) **
510   REM   * VARIABLES: I, N, T$, X$, Y$ *
520   DIM X(25), Y(25)

530   REM   TITLE OF DATA
540   READ T$
550   REM   NAME OF X-VARIATE
560   READ X$
570   REM   NAME OF Y-VARIATE
580   READ Y$
590   REM   NUMBER OF UNITS
600   READ N

610      FOR I = 1 TO N
620      READ X(I), Y(I)
630      NEXT I

640   REM   * OUTPUT: DATA X(·), Y(·); TITLE OF DATA T$ *
650   REM   * NAMES OF VARIATES X$, Y$; NUMBER OF UNITS N
```

Algorithm 6.5 *Input – Bivariate Data (Small Set)*

a man is a unit and the two variates are height and weight. For small sets of data, say less than 20 units (pairs), the data might be input using DATA statements; a check is made by reading over the DATA statements. Algorithm 6.5, *Input – Bivariate Data (Small Set)*, is appropriate for this simple type of input.

The input of larger sets of data should incorporate checks as the data are being entered. An algorithm of a similar form to *Input – Data Array* (Algorithm 4.1) would be suitable: we suggest it should be called *Input – Bivariate Data*. An illustration of how the control part of the program might run is as follows.

```
UNIT    NUMBER? 8
X?      12.1
Y?      29.8

THE NUMBERS ENTERED WERE
UNIT    X       Y
5       13.9    36.4
6       11.4    28.2
7       13.1    34.0
8       12.1    29.8

IF CORRECT TYPE Y
OTHERWISE TYPE N
```

6.6 Calculation of product-moment correlation coefficient

To obtain the correlation coefficient r is a one-line calculation when we have the results of an algorithm (6.3 or 6.4) for finding sums of squares and products (*ABC*, Section 20.3):

```
700   LET R = S(1,2)/SQR(S(1,1)*S(2,2))
```

It is always wise, before calculating a correlation coefficient, to plot a scatter diagram to check whether the points representing the data pairs (x_i, y_i) are approximately linearly placed, or whether there is any systematic curved trend. We recommend that the reader should always do this; a method of incorporating the plot and the calculation of the correlation coefficient in the same program is suggested in Section 6.8.

6.7 Calculation of Spearman's rank correlation coefficient

This calculation is made from N pairs of ranks. For example, a teacher might rank his class of pupils according to (i) their mathematical ability and (ii) their musical ability. He might assign ranks from his general knowledge of the pupils, or by putting in order the marks from examinations in mathematics and music. The calculation is exactly the same as for the product-moment correlation, with the ranks of one variate (e.g. mathematical ability) put in $X(\cdot)$ and the ranks of the other variate (e.g. musical ability) put in $Y(\cdot)$, (*ABC*, Section 20.9).

6.8 The menu method of assembling Algorithms

Analysing data is not a straightforward procedure like solving a quadratic equation. Data are investigated in various ways; the tools used are tables, graphs, summary

```
10   REM  ** BIVARIATE SAMPLE ANALYSIS **
20   REM  * VARIABLES: A$, I, R *

30   PRINT "DO YOU WANT THE MENU?"
40   PRINT "IF YES - TYPE Y AND PRESS RETURN"
50   PRINT "IF NO - PRESS RETURN"
60   INPUT A$
70   IF A$ <> "Y" THEN 3020

80   PRINT "1. INPUT OF DATA"
90   PRINT
100  PRINT "2. SCATTERGRAM"
110  PRINT
120  PRINT "3. CORRELATION COEFFICIENT"
130  PRINT
140  PRINT "TYPE NUMBER OF ITEM REQUIRED AND PRESS RETURN"
150  INPUT I

400  IF I <> 1 THEN 1400

500  REM  ** INPUT - BIVARIATE DATA (SMALL SET) **
510  REM  * VARIABLES: I, N, T$, X$, Y$ *
520  DIM X(25), Y(25)
···  (Algorithm 6.5)
640  REM  * OUTPUT: DATA X(•), Y(•); NUMBER OF UNITS N *
650  REM  * TITLE OF DATA T$, NAMES OF VARIATES X$, Y$ *
660  GO TO 30

1400 IF I <> 2 THEN 2400

1500 REM  ** OUTPUT - SCATTERGRAM **
1510 REM  * INPUT: NUMBER OF UNITS N *
1520 REM  * DATA X(•) POSITIVE AND IN ASCENDING ORDER *
1530 REM  * DATA Y(•) POSITIVE *
1540 REM  * TITLE T$, NAMES OF VARIATES X$, Y$ *
1550 DIM A(25), B(25)
···· (Algorithm 5.5)
1960 GO TO 30

2400 IF I <> 3 THEN 3000

2500 REM  ** SS AND SP (DEVIATION METHOD) **
2510 REM  * INPUT: DATA X(•), Y(•); NUMBER OF UNITS N *
2520 REM  * VARIABLES: I, M1, M2, T1, T2, T3, W1, W2 *
2530 DIM S(2,2)
···· (Algorithm 6.3)
2780 REM  * OUTPUT: MEANS M1, M2, SS AND SP MATRIX S(•,•) *

2790 LET R = S(1,2)/SQR(S(1,1)*S(2,2))
2800 PRINT "BETWEEN "; X$; " AND "; Y$
2810 PRINT "CORRELATION COEFFICIENT IS "; R
2820 GO TO 30

3000 PRINT "NUMBER NOT RECOGNISED - REPEAT"
3010 GO TO 80
3020 END
```

Program 6.6 *Bivariate Sample Analysis*

statistics and statistical tests. The outcome of one procedure may influence which other procedure is used next. The data may be modified, as the analysis proceeds, by rejecting some items or by the use of a mathematical transformation. Thus a computer program in which there is a fixed sequence of algorithms is not likely to be convenient for a complete statistical analysis. We here suggest a form of program that allows an interactive type of analysis.

The essence of the method is that the program offers a list of procedures, commonly called a **menu**, of which one is chosen, carried out, and followed by a return to the menu. Each time the menu is consulted another procedure may be chosen or the analysis stopped. In the simple example we present in Program 6.6, *Bivariate Sample Analysis*, three procedures are offered: (1) input of bivariate data; (2) plot of a scattergram; (3) calculation of a correlation coefficient. (A professional analysis of bivariate data would need many more procedures.) The program should be self-explanatory. The menu is obvious in lines 80–140. Choice of Option 1, in response to the input instruction of line 150, leads to entry into the input routine beginning in line 500; choice of Option 2 leads to the plotting routine beginning in line 1500; etc. Algorithms taken from other parts of the book begin at lines $500 + 1000k$ (for $k = 0, 1, 2$) to make re-numbering easy.

If a large set of data is to be worked over many times it will be worth putting the data into a data file on tape or disk. Procedures for setting up files are not part of the BASIC language and depend very much on the machine being used. We do not therefore go into details; the interested reader should look at the manual of the machine he is using.

6.9 Test data for Algorithms

Algorithm 6.1, *Mean and Variance (Deviation Method)*. Input $N = 5$, $X(\cdot) = (1, 2, 3, 5, 9)$. Output: $M = 4$, $V = 10$.
Algorithm 6.2, *Mean and Variance (Update Method)*. The same as for Algorithm 6.1 above.
Algorithm 6.3, *SS and SP (Deviation Method)*. Input: $N = 5$, Data pairs (x, y): $(1, 3), (2, 7), (3, 11), (5, 19), (9, 5)$. Output: $M1 = 4$, $M2 = 9$, $S(1, 1) = 40$, $S(1, 2) = 10$, $S(2, 2) = 160$.
Algorithm 6.4, *SS and SP (Update Method)*. The same as for Algorithm 6.3 above.

6.10 Exercises

1 Using Algorithm 6.1, *Mean and Variance (Deviation Method)*, construct a program which will print out the mean and the standard error of the mean, which equals SQR(V/N), of as many sets of data as required, using (i) interactive mode, (ii) batch mode. Test the program with the following data sets.
1) Number of children in family: 3, 1, 1, 2, 7.
2) Reaction time (in milliseconds): 217, 161, 198, 242, 140.
3) Age (in years): 19.5, 19.7, 20.1, 21.1, 22.6.

2 Run Algorithm 6.1, *Mean and Variance (Deviation Method)*, with data sets of the following form:

$$1 + 10^{-k}, 1 + 2 \times 10^{-k}, 1 + 3 \times 10^{-k}, 1 + 5 \times 10^{-k}, 1 + 9 \times 10^{-k}$$

in which k takes the values 1, 2, 3, ... Thus the first three data sets are:
1) 1.1, 1.2, 1.3, 1.5, 1.9;
2) 1.01, 1.02, 1.03, 1.05, 1.09;
3) 1.001, 1.002, 1.003, 1.005, 1.009.

The exact values of mean and estimated variance are, respectively, $1 + 4 \times 10^{-k}$ and 10^{1-2k}. Note the first value of k at which inexact values are printed, and continue until ridiculous values are obtained.

3   Write a program to calculate and print the mean and estimated variance of a set of data, using formula (2) of Section 6.1 to calculate the variance. Test this program using the data sets described in question 2.

4   In the program of question 1, print the standard error to F9 significant figures, and print the mean to the same number of *decimal places* as the standard error (use FØ for the number of decimal places). For most data, F9 = 2 will be suitable; for very accurately recorded data put F9 = 3.

5   In quality control, where observations are collected at fixed (equal) time intervals, the random component of variation is often estimated by

$$s^2 = \frac{1}{2(N-1)} \sum_{i=1}^{N-1} (x_{i+1} - x_i)^2$$

where x_i is the observation recorded at time i, and $i = 1, 2, \ldots N$. Write a program to calculate s^2, and to keep it updated as fresh observations are recorded.

6   Write a program to calculate the correlation coefficient between X and Y given N pairs of data (x, y), using either Algorithm 6.3 or 6.4.
   Use the program to calculate the correlation coefficient from N pairs of data with $x = i$, $y = i(10 - i)$, and $i = 1$ to N. Print out the values of the coefficient as N takes the values 4, 5, 6, 7, 8, 9, and note the influence of the number of pairs on the value of the coefficient.

7†   *Spearman's Rank Correlation Coefficient.* There are two methods of carrying out this calculation; one is described in Section 6.7 and the other in *ABC*, Section 20.9. The methods of constructing programs to calculate the coefficients are as follows.

(A) (based on Section 6.7).
   (1) Input the number of units N, and the paired data arrays X(·), Y(·), each of N elements.
   (2) Calculate the ranks of X(·) using *Ranking* (Algorithm 3.3) as a subroutine, and put the ranks in X(·).
   (3) Repeat (2) on Y(·).
   (4) Calculate the SS and SP matrix S(· , ·), using either Algorithm 6.3 or 6.4 as a subroutine.
   (5) Print the rank correlation coefficient

   S(1, 2)/SQR(S(1, 1)*S(2, 2)).

(B) (based on the formula $1 - 6\sum d_i^2 / n(n^2 - 1)$, where d_i is the difference in the ranks of X, Y for unit i; see *ABC*, Section 20.9).
  (1), (2), (3) as in (A) above.
  (4) Calculate $T = \sum d_i^2$, where $d_i = X(I) - Y(I)$, the difference of the *ranks* of the two variables.
  (5) Print the rank correlation coefficient

$$1 - 6*T/(N*(N*N - 1)).$$

Write a program for each of these methods, and use it on the data in *ABC*, Section 20.9, on the rankings given to strawberry jam samples by two judges:

Sample	i	ii	iii	iv	v	vi	vii	viii	ix	x
Judge A's rank (X)	7	8	1	6	3	9	2	4	5	10
Judge B's rank (Y)	10	9	4	3	6	8	1	2	5	7

8 Write a program to calculate the mean and variance of a set of data which are given in the form of a grouped frequency table. Assume that the ends of each interval are given; calculate the mid-points $X(I)$ of the intervals, and let $F(I)$ be the frequency of observations in interval I, with I taking values 1 to N. Then the mean and variance are found from the formulae (*ABC*, Section 4.5.2):

$$\bar{x} = \sum_{i=1}^{n} f_i x_i \bigg/ \sum_{i=1}^{n} f_i$$

and $\quad s^2 = \sum_{i=1}^{n} f_i(x_i - \bar{x})^2 \bigg/ \left(\sum_{i=1}^{n} f_i - 1 \right).$

The formula for s^2 is the extension of formula (1) in Section 6.1. Your program can therefore be regarded as the extension of Algorithm 6.1.

9 Write flow diagrams for Algorithms 6.3 and 6.4.

7 Simulation

7.1 Introduction

A favourite game on computers is 'lunar lander'. In one version there is a lunar landscape displayed at the bottom of a VDU screen and a spacecraft appears at the top of the screen. The aim of the player is to make a gentle landing of the spacecraft on the moon's surface. Control of the spacecraft's movement is exercised by modifying the thrust of the two engines of the spacecraft; the player is able to input this information from the keyboard. The program in the computer calculates the position at time $t + 1$ given the thrust of the engines, and the position and velocity of the spacecraft at time t. The motion of a real spacecraft is thus mimicked by the computer; we call this **simulation**. In a simulation we pretend something is happening and work through the essential stages of the process, often using a mathematical model.

Computers make simulation so easy that it is a technique widely used in, for example, industry, business, teaching and research. We give only a brief introduction to the subject in this chapter and in Chapter 9; it may be pursued in Tocher (1963), Fishman (1978) and Yakowitz (1977).

The 'lunar lander' game described above is a **deterministic** simulation. When we know the state of the spacecraft (i.e. its position and velocity and the thrust of its engines) at time t, then its state at time $t + 1$ is completely determined; no chance or random element is involved in the program. We shall now consider in detail a **stochastic** simulation, where a chance element is involved. A large part of the chapter is given over to discussing the generation of random numbers, the means by which the chance element is brought into a simulation.

7.2 Simulation of card collecting

Cigarette manufacturers used to give away cards in packets of cigarettes. Each card was from a set whose subject might be famous cricketers or film stars or breeds of dogs. A desire to complete the set led smokers to buy more cigarettes; other manufacturers use the same idea today. Using a computer simulation, it is easy to mimic how a collection would be built up. Probably the question of most interest is: 'What is the distribution of the number of packets that need to be bought to complete a set of cards?' This may be answered easily from the simulation. The core of the simulation is shown in a flow diagram in Figure 7.1. A program for the simulation is given in Program 7.1 (page 66).

Suppose that there are N1 cards in a complete set. The state of the collector's set at any time is stored in a 'wants' array W$(·) with N1 elements; initially each element is set to 'WANTED' and when the Ith card is collected W$(I) is changed to 'PRESENT'. A variable C$ is set initially to 'INCOMPLETE' and is changed to 'COMPLETE' when each element of W$(·) equals 'PRESENT'.

Collecting a card is represented by choosing a random integer between 1 and N1

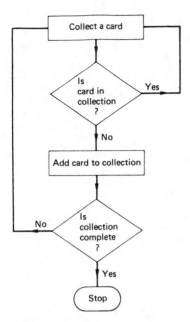

Figure 7.1

inclusive. Most forms of BASIC include a random number generator. Thus the statement

```
51Ø    LET W = RND(1)
```

puts W equal to a random number between 0 and 1, and the value is intended to be a random sample from the continuous uniform distribution (*ABC*, Section 13.5) on the interval (0, 1). (The generation of random numbers by computer is discussed in Section 7.3.) Each time the instruction is obeyed a new random sample is chosen from (0, 1). The value '1' in RND(1) has no special significance; in most versions of BASIC any value greater than zero would give the same result, whereas use of a negative argument allows the repetition of a sequence of random numbers and is useful when checking programs. (The reader should check the details of the RND function on his own machine.) The instructions on lines 13Ø and 14Ø use the RND function to produce a random integer between 1 and N1 inclusive.

The total number of cards inspected to complete the set (which equals the number of packets purchased) is counted in N and this is printed. The number of cards of each sort is counted in the N1 elements of the array H(·); this information may be printed if required. Ten runs of the program with 30 cards in a set gave the number of packets N as:

116, 106, 213, 94, 156, 124, 206, 125, 102, 96.

These numbers illustrate a common characteristic of stochastic simulations: the results are often very variable.

7.3 Generation of pseudo-random numbers

We expect the reader to use the RND function that should be available on his machine to generate the random numbers he requires for simulations. But random number

```
10   REM  ** CARD COLLECTING SIMULATION **
20   REM  * VARIABLES: A$, C$, I, J, N, N1 *
30   DIM H(50), W$(50)

40   PRINT "NUMBER IN SET";
50   INPUT N1

60   LET C$ = "INCOMPLETE"                          Initialise arrays
70      FOR I = 1 TO N1                             and counters
80      LET W$(I) = "WANTED"
90      LET H(I) = 0
100     NEXT I
110  LET N = 0

120  REM  COLLECT CARD
130  LET I = N1*RND(1)                              Identify card
140  LET I = INT(I) + 1
150  LET H(I) = H(I) + 1
160  LET N = N + 1

170  REM  TEST IF CARD IN COLLECTION
180  IF W$(I) = "WANTED" THEN 200
190  GO TO 120
200  LET W$(I) = "PRESENT"

210  REM  TEST IF COLLECTION COMPLETE
220     FOR J = 1 TO N1
230     IF W$(J) = "WANTED" THEN 260
240     NEXT J
250  LET C$ = "COMPLETE"
260  IF C$ = "INCOMPLETE" THEN 120
270  PRINT N; "    PACKETS PURCHASED"
280  PRINT
290  PRINT "RERUN WITH SAME SET SIZE (Y OR N)";
300  INPUT A$
310  PRINT
320  IF A$ = "Y" THEN 60
330  END
```

Program 7.1 *Card Collecting Simulation*

generators are not perfect so it is wise to understand the methods that are used and to appreciate their limitations.

Before computers became widely available, statisticians used tables of random digits to choose random samples from populations or distributions. A line of such a table might take the form:

17 53 77 58 71   71 41 61 50 72   12 41 94 96 26   44 95 27 36 99

One might consider storing a table of random digits in a computer but currently no-one does so. To be speedily available they would have to be stored in the central part of the computer, and programmers consider that to be a waste of precious storage space. If, alternatively, the numbers were put in an ancillary storage device such as a disk or cassette, retrieval would be too slow. What is desired is that the computer shall produce lots of random numbers on demand, using a simple program that takes up little storage space. The successive digits in the decimal expansion of π come to mind as digits that have no obvious pattern and that we might therefore consider 'random', although their generation would take too long for it to be a convenient method of obtaining random digits, even if it satisfied the other necessary criteria. A moment's thought makes us ask whether we are demanding the impossible. We want the numbers to be generated by a short program and hence by a completely deterministic process. How can they possibly be *random* numbers? We have to admit that they cannot be random numbers in any strict sense (although investigation has shown that it is difficult to say exactly what a random number is) and that all the current random number generators have faults. The aim is to produce numbers that have the properties we expect random numbers to have, if they are to be satisfactory in the application we have in mind. A more correct term for the numbers produced is **pseudo-random** numbers, though for brevity we shall refer to them as random numbers.

For non-computer use, statisticians found random *digits* the most convenient form of random number. The numbers were random samples from a discrete uniform distribution (*ABC*, Section 7.4) taking values 0, 1, 2, . . . , 9 each with probability 0.1. The digits were combined as required. The most convenient random numbers for computer use, however, are random samples from the continuous uniform distribution (*ABC*, Section 13.5) on the interval (0, 1). We may represent this distribution by U(0, 1). The values are of course truncated since the computer can hold numbers only to a fixed number of places. Examples of pseudo-random numbers generated by a microcomputer are:

.131 137 465, .809 248 730, .846 447 204, .841 536 558, .591 965 711, .268 001 130.

Random digits, if required, are generated by the instructions:

```
500    LET W = RND(1)
510    LET W = INT(10*W)
```

We shall see later (Chapter 9) that random samples from U(0, 1) are the basis for obtaining random samples from any statistical distribution.

To sum up our requirements for the pseudo-random numbers: we wish to generate efficiently (both in speed and storage) numbers that we may regard as independent (an alternative but more precise word than 'random') samples from U(0, 1).

7.3.1 The mid-square method

One of the first approaches to this problem was that of John von Neumann, a mathematician who made many early contributions to the theory of computing. A p-digit number w_0 is squared. The central p digits of w_0^2 are taken as w_1, which in turn is squared and the resulting central digits taken as w_2, etc. With $p = 4$, we might begin with $w_0 = 7137$ to give $w_0^2 = 50\,936\,769$, yielding $w_1 = 9367$. The sequence obtained, after dividing each number by 10^4 to obtain values in the interval (0, 1), is

.7137, .9367, .7406, .8488, .0461, .1215, .5156, . . .

There are 10 000 possible values of w (using $p = 4$). Hence at some stage a number will repeat, and from then on the whole sequence will repeat. The maximum possible length of a cycle would be 10 000 but cycles are in practice much shorter. For example, with $w_0 = 1357$ we obtain the following sequence of 4-digit numbers:

$$1357, \ 8414, \ 7953, \ 2502, \ 2600, \ 7600, \ 7600, \ 7600, \ \ldots .$$

The sequence settles into a cycle of period 1. Another sequence is

$$9999, \ 9800, \ 0400, \ 1600, \ 5600, \ 3600, \ 9600, \ 1600, \ \ldots$$

which yields a cycle of period 4. These features are not satisfactory. Moreover it can be shown (Tocher, 1963) that the 10^p possible values are not equally likely; therefore the method fails in one of the basic requirements of a random sequence. But it does suggest a way of proceeding. We seek to generate a sequence $w_i (i = 0, 1, 2, \ldots)$ where each term is generated from the previous one by a relation $w_{i+1} = f(w_i)$ such that the cycle length is very long.

7.4   The congruence method

Another mathematician, Lehmer, suggested using the theory of numbers to devise the required sequences; in particular, using the theory of congruences. The generating relation is

$$w_{i+1} = a_1 w_i + a_2 \quad (\text{mod } k)$$

where all the quantities are integers. The right-hand side $(a_1 w_i + a_2)$ is evaluated, and working modulo k means that the expression is replaced by the remainder after dividing by k. Let us use simple numbers ($a_1 = 21$, $a_2 = 3$, $k = 100$) to illustrate the idea; we might begin with w_0 (often called the **seed**) equal to 7. Then

$$w_1 = 21 \times 7 + 3 \ (\text{mod } 100) = 147 + 3 \ (\text{mod } 100) = 150 \ (\text{mod } 100) = 50.$$
$$w_2 = 21 \times 50 + 3 \ (\text{mod } 100) = 1050 + 3 \ (\text{mod } 100) = 1053 \ (\text{mod } 100) = 53.$$

Division by the modulus 100 puts the numbers in the interval (0, 1) and we obtain the sequence: .07, .50, .53, .16, .39, ...

We can outline an algorithm to generate random numbers:

(1) input seed w_0, an integer;
(2) $w_1 = a_1 w_0 + a_2 \ (\text{mod } k)$; the result is an integer;
(3) $u = w_1/k$, where the result is truncated to a given number of places;
(4) output u, a sample from U(0, 1).

The choice of a_1, a_2 and k is critical for the performance of the generators. Two general classes of generators, depending on whether $a_2 = 0$ or not, may be recognised. The first congruential generators were of the form

$$w_{i+1} = a_1 w_i \quad (\text{mod } k)$$

and, for obvious reasons, are called **multiplicative** congruential generators. The generators containing the additional positive integer a_2 are called **mixed** congruential generators. When choosing the parameters of a generator the properties that can be taken account of are (a) efficiency, (b) long cycle length, and (c) small serial correlation.

Here efficiency means quick generation. Steps (2) and (3) of the algorithm above involve division, which is usually a relatively slow operation on a computer. Therefore it is desirable to choose the modulus k to make division by it quick and simple. This can be achieved if k is a power of the number base b of the computer (most commonly $b = 2$, but it may equal 10). The parallel in ordinary arithmetic is when, working with base 10, the division of 179236 by 10^3 is accomplished simply by moving a decimal point, to give 179.236. (Division by $b^p \pm 1$ is also achieved simply on a computer with number base b, by subtracting the appropriate number from the word (or bytes) containing the exponent.) We also want k to be as large as possible so that the maximum cycle is long, and hence efficiency suggests that the modulus k should equal b^l (or $b^l \pm 1$), where b is the number base of the computer and l is related to the word length of the computer. (For the common microcomputers an appropriate value of l is 15; with mainframe computers it will be about twice this.) We assume, of course, that the random number generator will be programmed in machine code.

Rules for the choice of parameters to give the maximum possible cycle length with mixed congruential generators have been found. A sequence from a mixed congruential generator has cycle length k if and only if:

(i) a_2 and k have no common factors except 1;
(ii) $(a_1 - 1)$ is a multiple of every prime that divides k;
(iii) $(a_1 - 1)$ is a multiple of 4 if k is a multiple of 4.

It is easy to see that, if k is chosen as recommended in the previous paragraph, then to obey these rules we should choose a_1 equal to $20c + 1$ (c being a positive integer) on a decimal machine, and equal to $4c + 1$ on a binary machine.

Maximum possible cycle lengths are not obtainable with multiplicative congruential generators, though long ones can be found. For example, if $k = 2^\alpha$ then a cycle length of $2^{\alpha - 2}$ may be obtained. With these generators, the seed w_0 must be chosen so that it has no factors in common with the modulus k.

In theory, the sample values obtained are required to be independent, so the serial correlations will be zero. The most important requirement in practice is that the first serial correlation (the correlation of u_i with u_{i+1}) shall be small, for we cannot ensure that it is zero. It has been shown that with a mixed congruential generator in which a_2 is much less than k, this requirement can be satisfied by choosing a_1 approximately equal to $\sqrt{k}$.

Unfortunately, satisfying these desirable properties does not ensure that we have a good random number generator. The only way to check whether a generator is producing suitably 'random' numbers is to apply to the numbers the tests described in Section 7.6; but first we will give an example of a random number generator.

7.5   Example of a linear congruential random number generator

We quote an example from Yakowitz (1977) to illustrate the way a random number generator works. We are working in BASIC, which effectively uses a number base of 10, so it is convenient to make the modulus a power of 10. (We stress that this is an illustration and that it produces a slow generator. The random number generator on your machine most probably uses binary arithmetic and is written in machine code.)

The actual values are $k = 10^5, a_1 = 1001, a_2 = 457$. These parameters satisfy the rules given in the previous section and therefore give the maximum possible cycle length of 10^5.

The operation of finding the value of an integer $w \pmod{k}$, which is Step (2) in the algorithm of Section 7.4, may be carried out using the expression

$$[(w/k) - \text{INT}(w/k)] * k$$

provided that the arithmetic is done exactly. (The expression is easily verified using $w = 63$ and $k = 10$.) Since Step (3) is division by k, the expression

$$[(w/k) - \text{INT}(w/k)]$$

carries out Steps (2) and (3) and produces the required pseudo-random number u. Finally, note that since

$$w_1/k = a_1(w_0/k) + (a_2/k)$$

we may write the linear relation in the form

$$u_1 = 1001 \, u_0 + 0.00457.$$

```
500   REM   ** RANDOM **
510   REM   * INPUT: SEED U0 *
520   REM   * VARIABLES: A1, A2 *

530   LET A1 = 1001
540   LET A2 = 0.00457

550   LET U0 = A1*U0 + A2
560   LET U0 = U0 - INT(U0)

570   LET U0 = INT(U0*100000 + 0.5)/100000

580   REM   * OUTPUT: SAMPLE U0 FROM U(0,1) *
```

Algorithm 7.2 *Random*

We made a proviso above that the arithmetic be exact. Computers cannot be relied upon to do exact arithmetic. If 5 decimal places is well within the accuracy of the machine being used, then line 570 of Algorithm 7.2, *Random*, ensures that the arithmetic is kept exact. The reader will probably find it instructive to compare repeated use of the algorithm with and without line 570.

7.6   Tests of randomness

The only reasonable guarantee we can have that a pseudo-random number generator will be satisfactory is that it has passed tests of the statistical properties of the numbers it produces. These properties may be summed up as the requirements of **uniformity** and **independence**. Many statistical tests have been used on random numbers; we have space for a few only. More tests are given in Fishman (1978), for example.

7.6.1 Frequency test

This is a test of uniformity. If the (0, 1) interval is divided into a number of classes of equal width, then we would expect an equal number of random numbers to fall in each class. For example, if we use 10 intervals each of length 0.1 then, with 1000 random numbers, we would expect 100 to fall in each of the ten intervals (classes). The frequencies of the numbers falling in each class may be found using Algorithm 4.4, *Frequency Table*. The chi-squared test of equal frequencies may be done using Algorithm 10.6, *Chi-squared Goodness-of-fit*.

7.6.2 Serial test

We suppose the (0, 1) interval is divided into classes as in the frequency test. We then count the frequencies $F(I, J)$ of pairs of random numbers such that the first of the pair falls in interval I and the second falls in interval J. These frequencies should all be equal if the numbers are uniform and independent. The chi-squared test may again be carried out with Algorithm 10.6, *Chi-squared Goodness-of-fit*.

7.6.3 Serial correlation

The first serial correlation of the sequence $u_0, u_1, \ldots u_N$ is the correlation of u_i with u_{i+1}, where i runs through the values $0, 1, 2, \ldots N-1$. The correlation coefficient may be calculated as described in Section 6.6, and tested for difference from zero as described in Exercises 10.7, question 2. The array $X(\cdot)$ used in the algorithms will have elements $u_0, u_1, \ldots u_{N-1}$ while the array $Y(\cdot)$ will have elements $u_1, u_2, \ldots u_N$. Similarly, the tth serial correlation coefficient measures the correlation of u_i with u_{i+t}. These serial correlations, for all values of t, should be zero if the numbers being produced are independent.

[*Note*: We give no test data for the algorithms in this chapter, but some of the following exercises will serve this purpose.]

7.7 Exercises

1 Run Program 7.1, *Card Collecting Simulation*, with N1 = 10. Repeat this a number of times, and observe the variation in the number of packets that must be bought to complete a set of cards.

Print the array $H(\cdot)$, which contains the number of cards of each of the N1 types that are collected in order to obtain a complete set.

2 It is necessary to run *Card Collecting Simulation* (Program 7.1) many times in order to investigate the distribution of the number of packets that need to be bought in order to complete a set. Modify the program so that it will carry out N2 successive simulations, without interruption, and print out the frequency distribution of the number of packets purchased.

Run the program with the number of cards in the set (N1) equal to 10, and the number of simulations (N2) equal to 100. [Note that for many sets of cards published recently, N1 = 40, or perhaps 30, would be more realistic: the number of simulations that can conveniently be done for these values of N1 may have to be reduced.]

3 Continuing question 2, we may call the interval between a card being obtained and another wanted card appearing a *waiting time*. In the simulation this is measured as number of packets. Extend your program so that in any given simulation the waiting time to the ith card is stored in $W(I)$; $W(1)$ will of course be 1. For each I (from 1 to N1), total the waiting times $W(\cdot)$ for the N2 simulations; store these totals in $T(\cdot)$. Then $T(I)/N2$ will be the mean waiting time for card I; print this out for each I in addition to the other quantities printed by the program.

4 Generate 100 random numbers using Algorithm 7.2, *Random*. By using or adapting *Frequency Table* (Algorithm 4.4) and *Output – Frequency Table* (Algorithm 4.5), print out a frequency table using classes with lower endpoint 0 and class-width 0.1.

5 *Frequency test on pseudo-random numbers*. Construct a program to generate and test random numbers according to the description given in Section 7.6.1.

6 *Serial test on pseudo-random numbers*. Construct a program to generate and test random numbers according to the description given in Section 7.6.2.

7 *Serial correlation in pseudo-random numbers*. Construct a program to generate random numbers, and to test them by calculating the first, second and third serial correlation coefficients (i.e. $t = 1, 2$ and 3) according to the description given in Section 7.6.3.

8 Write a flow diagram for *Card Collecting Simulation*, Program 7.1, as amended in questions 2 and 3 above.

8 Probability functions of random variables

8.1 Introduction

There are two functions associated with each discrete random variable X: the **probability mass function** and the **cumulative distribution function** (*ABC*, Chapter 7). If we denote the probability mass function by the letter f, then the probability that X takes the particular value x is.

$$\Pr(X = x) = f(x).$$

For example, in the binomial distribution having parameters n and p,

$$f(x) = \binom{n}{x} p^x (1-p)^{n-x}$$

where x may take any of the values 0, 1, 2, . . . n.

The letter F is used to denote the cumulative distribution function, which gives the probability that the random variable X is less than or equal to some particular value, say b. Thus

$$F(b) = \Pr(X \le b).$$

For a random variable taking values 0, 1, 2, . . . n, such as the binomial,

$$F(b) = \sum_{x=0}^{b} f(x) \qquad (b \le n).$$

Similarly, two functions are defined for each continuous random variable X (*ABC*, Chapter 13). The **probability density function** $f(x)$ of X is a function whose integral from $x = a$ to $x = b$ ($b \ge a$) gives the probability that X takes a value in the interval (a, b):

$$\Pr(a \le X \le b) = \int_a^b f(x)\mathrm{d}x.$$

The **cumulative distribution function** $F(b)$ is such that

$$F(b) = \Pr(X \le b) = \int_{-\infty}^b f(x)\mathrm{d}x.$$

With the exponential distribution, for example, the probability density function is

$$f(x) = \lambda e^{-\lambda x} \qquad (x \ge 0),$$

and its cumulative distribution function is

$$F(b) = \int_0^b \lambda e^{-\lambda x}\mathrm{d}x = 1 - e^{-\lambda b} \qquad (b \ge 0).$$

In this chapter we consider how to calculate the values of $f(x)$ and $F(b)$ for common statistical distributions.

8.2   Discrete random variables

8.2.1   Binomial distribution

The probability of x successes in n independent trials, when the probability of success at any trial is p, is given by

$$f(x) = \binom{n}{x} p^x q^{n-x} \qquad (x = 0, 1, \ldots n),$$

where $q = 1 - p$. Algorithm 8.1, *Binomial Probabilities*, is based on the recurrence formula relating the probability of i successes $P(i)$ to the probability of $i - 1$ successes, i.e.

$$P(i) = \frac{(n - i + 1)}{i} \frac{p}{q} P(i - 1),$$

with the starting value $P(0) = q^n$. The algorithm generates the probabilities of all possible values of the random variable.

```
5ØØ  REM  ** BINOMIAL PROBABILITIES **
51Ø  REM  * INPUT: NUMBER OF TRIALS N *
52Ø  REM  * PROBABILITY OF SUCCESS AT ANY TRIAL P *
53Ø  REM  * VARIABLES: I, N1, Q, W *
54Ø  DIM P(25)

55Ø  LET Q = 1 - P
56Ø  LET W = Q↑N
57Ø  LET P(Ø) = W
58Ø  LET N1 = N + 1
59Ø      FOR I = 1 TO N
6ØØ      LET W = (N1 - I)*P*W/(I*Q)
61Ø      LET P(I) = W
62Ø      NEXT I

63Ø  REM  * OUTPUT: BINOMIAL PROBABILITIES P(·) *
```

Algorithm 8.1 *Binomial Probabilities*

The Jth element $(J = 0, 1, \ldots N)$ of the array giving the values of the cumulative distribution function is $\sum\limits_{i=0}^{J} P(i)$.

8.2.2   Poisson distribution

The probability mass function of the Poisson distribution with parameter λ is given (*ABC*, Chapter 19) by

$$f(x) = \frac{\lambda^x}{x!} e^{-\lambda} \qquad (x = 0, 1, 2, \ldots).$$

The calculation of the probabilities may be carried out in a similar way to that for the binomial distribution (Algorithm 8.1), but based on the recurrence relation

$$P(i) = \lambda P(i-1)/i \qquad (i = 1, 2, 3, \ldots)$$

where $P(0) = e^{-\lambda}$.

The possible values of the Poisson random variable do not have a finite upper limit, as do those of the binomial. However, unless λ is very large (which is most unusual), it is only the first few values of $P(i)$ that we shall need. It is therefore necessary to include as input to the algorithm the largest value, i, of the variable that is of interest: call this U1. The calculations of probabilities are then carried out up to U1, and the summations of the $P(i)$ to obtain the cumulative distribution function also end at U1. We recommend using the variable L in the program to represent the parameter λ.

8.2.3 Geometric distribution

The probability mass function of the geometric distribution with parameter p is given (*ABC*, Section 7.7) by

$$f(x) = p(1-p)^{x-1} \qquad (x = 1, 2, \ldots).$$

An appropriate recurrence relation is

$$P(i) = (1-p)P(i-1),$$

i.e. $\qquad P(i) = qP(i-1),$

where $q = 1 - p$, for $i = 2, 3, 4, \ldots$; and

$$P(1) = p.$$

Again it is easy to construct an algorithm similar to Algorithm 8.1 but based on this recurrence relation. We must include a specified U1 for the same reason as for the Poisson distribution (see Section 8.2.2 above).

8.3 The exponential distribution

The probability density function of the exponential distribution, which is a continuous random variable, is given (*ABC*, Section 13.6) by

$$f(x) = \lambda e^{-\lambda x} \qquad (x \geq 0),$$

where $\lambda > 0$. The distribution has mean $1/\lambda$. The cumulative distribution function is given by

$$F(b) = 1 - e^{-\lambda b}.$$

One line of program will serve to calculate each of these functions for a particular value of x (or b). This is an occasion when BASIC's facility for defining functions may be useful. Thus

```
10   DEF FNP(X) = L*EXP(-L*X)      Probability density function
20   DEF FNC(X) = 1 - EXP(-L*X)    Cumulative distribution function
```

Sometimes we wish to use the inverse distribution function; that is, to calculate the value of x corresponding to a particular value p of the cumulative distribution function. We therefore have

$$p = 1 - e^{-\lambda x}$$

whose solution is $x = -\lambda^{-1} \ln(1-p)$. The function we might define is

```
30    DEF FNI(P) = -(LOG(1 - P))/L    Inverse distribution function
```

8.4   Calculation of integrals

The probability density functions of most of the common continuous random variables look more complicated than that of the exponential distribution, but evaluation of them by computer is straightforward. Evaluating the cumulative distribution functions may be awkward since there is not always an equivalent explicit expression. The most direct approach in such cases is to use numerical integration of the probability density function. We therefore give Algorithm 8.2, *Simpson's Rule*, for integration using Simpson's rule. The probability density function being integrated is a defined function FNP(X).

```
1500   REM   ** SIMPSON'S RULE **
1510   REM   * INPUT: INTEGRAND FNP(X) *
1520   REM   * RANGE OF INTEGRATION (LØ,UØ); NUMBER OF STRIPS NØ *
1530   REM   * VARIABLES: D1, PØ, WØ, W1, W2 *

1540   LET D1 = (UØ - LØ)/NØ
1550   LET PØ = Ø
1560   LET WØ = FNP(LØ)
1570       FOR I = 1 TO NØ
1580       LET W1 = FNP(LØ + Ø.5*D1)
1590       LET LØ = LØ + D1
1600       LET W2 = FNP(LØ)
1610       LET PØ = PØ + D1*(WØ + 4*W1 + W2)/6
1620       LET WØ = W2
1630       NEXT I

1640   REM   * OUTPUT: INTEGRAL PØ *
```

Algorithm 8.2 *Simpson's Rule*

In evaluating a cumulative distribution function, the lower limit of the necessary integral is the lower limit of the possible values taken by the random variable: the range of integration begins at the left-hand end or tail of the distribution and moves up to the particular value b of interest. (Sometimes it is easier to find an integral related to this required one.) $F(b)$ is the probability that $X < b$, and we usually denote this cumulative distribution function by PØ in the algorithms.

The accuracy of the value of the integral we obtain by this method depends upon the number n of intervals (strips) into which the range of integration (a, b) is divided and the 'smoothness' of the integrand. Numerical analysis gives us a bound (upper limit) for the error which is

$$\varepsilon = \left| \frac{M(b-a)^5}{2800 \, n^4} \right|$$

where M is the maximum value of the modulus of the fourth derivative of the integrand in the interval. If we substitute for ε the accuracy within which we wish to work, it is easy to calculate n.

8.5   The normal integral

8.5.1   Simpson's rule

Algorithm 8.3, *Normal Distribution Fnc – Integration*, evaluates the normal cumulative distribution function for a chosen argument, using *Simpson's Rule* (Algorithm 8.2) as a subroutine. The integral to be evaluated (*ABC*, Section 14.2) is

$$F(z) = \int_{-\infty}^{z} \frac{1}{\sqrt{2\pi}} \exp\left(-\frac{t^2}{2}\right) dt.$$

We cannot integrate over an infinite range using Simpson's rule, so we must rewrite the integral; in doing so we make use of the fact that the integrand is symmetrical about $t = 0$. We obtain

$$F(z) = 0.5 + \int_{0}^{z} \frac{1}{\sqrt{2\pi}} \exp\left(-\frac{t^2}{2}\right) dt.$$

```
500   REM   ** NORMAL DISTRIBUTION FNC - INTEGRATION **
510   REM   * INPUT: ARGUMENT Z, NUMBER OF STRIPS NØ *
520   REM   * REQUIRES: SIMPSON'S RULE (8.2) AT 1500 *
530   REM   * VARIABLES: A1, LØ, UØ, PØ *

540   DEF FNP(X) = A1*EXP(-Ø.5*X*X)              Integrand
550   LET A1 = Ø.39894228                        1/√2π
560   LET LØ = Ø                                 Set limits of
570   LET UØ = Z                                 integration
580   GOSUB 1500                                 Simpson's rule
590   LET PØ = PØ + Ø.5

600   REM   * OUTPUT: PØ, PROBABILITY X < Z *
```

Algorithm 8.3 *Normal Distribution Fnc – Integration*

This is valid for positive and negative z. To fix an appropriate number, n, of strips we may use the formula for the error bound given in Section 8.4. The fourth derivative of $\exp(-t^2/2)$ is $(t^4 - 6t^2 + 3)\exp(-t^2/2)$, which has a maximum at $t = 0$ giving $M = 3/\sqrt{2\pi}$. For a chosen ε, we may now find n for our range of integration $(0, z)$. Remembering that n must be an integer, we find, if we choose $\varepsilon = 10^{-6}$, the following values of n for various z:

z	1	2	3	4
n	5	11	18	26

8.5.2 An approximation

An alternative approach is to use an approximation that is an explicit function of z. We give an approximation due to Hastings; it is quoted in Abramowitz and Stegun (1972), where more accurate—though more complicated—approximations are also quoted. For non-negative z,

$$F(z) = 1 - 0.5(1 + a_1 z + a_2 z^2 + a_3 z^3 + a_4 z^4)^{-4} + \varepsilon(z).$$

The values of the constants $a_1, \ldots a_4$ are given in Algorithm 8.4, *Normal Distribution Fnc – Approx*.

```
500  REM  ** NORMAL DISTRIBUTION FNC - APPROX **
510  REM  * INPUT: ARGUMENT Z *
520  REM  * VARIABLES: A1, A2, A3, A4, P0, W *

530  LET A1 = 0.196854
540  LET A2 = 0.115194
550  LET A3 = 0.000344
560  LET A4 = 0.019527

570  LET W = ABS(Z)
580  LET P0 = 1 + W*(A1+W*(A2+W*(A3+W*A4)))
590  LET P0 = P0↑4
600  LET P0 = 1 - 0.5/P0
610  LET P0 = 0.5 + (P0 - 0.5)*SGN(Z)

620  REM  * OUTPUT: P0, PROBABILITY X < Z *
```

Algorithm 8.4 *Normal Distribution Fnc – Approx*

For negative z we evaluate the expression using the absolute value of z and obtain the value P0. Then the probability we require is $1 - $ P0. This is dealt with in lines 570 and 610. The error of the approximation is $\varepsilon(z)$, and it is stated that $|\varepsilon(z)| < 2.5 \times 10^{-4}$.

8.5.3 Inverse normal integral by approximation

The cumulative distribution function $F(z)$ gives the probability that a value chosen at random from the distribution will be less than z. Sometimes we wish to carry out the inverse operation: given the probability, we wish to find the corresponding z. If we put $1 - F(z) = 1 - p = q$, then provided that $0 < q \le 0.5$ the z corresponding to a particular value of q is given by

$$z = t - \frac{a_1 + a_2 t}{1 + a_3 t + a_4 t^2} + \varepsilon(q),$$

where $t = \sqrt{(-2 \ln q)}$. The absolute error $|\varepsilon(q)| < 3 \times 10^{-3}$. The limitation on the range of q is removed if we put $Q = 0.5 - ABS(P - 0.5)$ and then take the answer as $Z*SGN(P - 0.5)$.

We give Algorithm 8.5, *Inverse Normal – Approx*, to carry out this calculation.

```
500  REM   ** INVERSE NORMAL - APPROX **
510  REM   * INPUT: PROBABILITY PØ *
520  REM   * VARIABLES: A1, A2, A3, A4, QØ, W, W1, W2, Z *

530  LET A1 = 2.30753
540  LET A2 = Ø.27Ø61
550  LET A3 = Ø.99229
560  LET A4 = Ø.Ø4481

570  LET QØ = Ø.5 - ABS(PØ - Ø.5)
580  LET W = SQR(-2*LOG(QØ))
590  LET W1 = A1 + A2*W
600  LET W2 = 1 + W*(A3 + W*A4)
610  LET Z = W - W1/W2
620  LET Z = Z*SGN(PØ - Ø.5)

630  REM   * OUTPUT: STANDARD NORMAL VALUE Z *
```

Algorithm 8.5 *Inverse Normal – Approx*

8.6*   Distributions based on the normal distribution

The sampling distributions t, χ^2 and F are important in both theoretical and practical statistics. Their probability density functions are as follows.

i) t with k degrees of freedom:

$$\left(\frac{\Gamma[(k+1)/2]}{\sqrt{k\pi}\,\Gamma[k/2]}\right)\left(\frac{1}{(1+x^2/k)^{(k+1)/2}}\right) \qquad -\infty < x < \infty.$$

ii) χ^2 with k degrees of freedom:

$$\frac{1}{\Gamma[k/2]2^{k/2}}\,x^{k/2-1}\,e^{-x/2} \qquad 0 < x < \infty.$$

iii) F with k_1 and k_2 degrees of freedom:

$$\left(\frac{\Gamma[(k_1+k_2)/2](k_1/k_2)^{k_1/2}}{\Gamma[k_1/2]\Gamma[k_2/2]}\right)\left(\frac{x^{(k_1-2)/2}}{(1+k_1x/k_2)^{(k_1+k_2)/2}}\right) \quad 0 < x < \infty.$$

These distributions are defined in terms of normal variables in Chapter 9, and their density functions are derived in Hogg and Craig (1970). The uses of t and χ^2 are discussed in *ABC* (Chapters 16 and following), and all three distributions are discussed in Snedecor and Cochran (1980).

The density functions involve the gamma function $\Gamma(x)$, which is closely related to factorials. The function $\Gamma(x)$ is equal to $(x-1)\Gamma(x-1)$, and hence if we put x equal to a positive integer n we obtain $\Gamma(n) = (n-1)!$ (since $\Gamma(1) = 1$). In the density functions above, the arguments are all such that twice the argument is a positive integer. Using the relation between $\Gamma(x)$ and $\Gamma(x-1)$, it follows that, if n is a positive integer, $\Gamma(n+\frac{1}{2})$ equals $(n-\frac{1}{2})(n-\frac{3}{2})(n-\frac{5}{2})\ldots\frac{3}{2}\cdot\frac{1}{2}\Gamma(\frac{1}{2})$; it can be shown also that $\Gamma(\frac{1}{2}) = \sqrt{\pi}$.

Now the gamma function can be dangerous to use on a computer since it very quickly overflows a small machine: an argument of 35 or 36 saturates it. We therefore work with $\ln \Gamma(x)$ instead, since we shall find this easy to incorporate into later algorithms. Algorithm 8.6, *Log Gamma Function*, evaluates the logarithm of the function for arguments of the form W, where 2W is a positive integer.

```
500   REM   ** LOG GAMMA FUNCTION **
510   REM   * INPUT: ARGUMENT W *
520   REM   * (2*W) IS AN INTEGER > 0 *
530   REM   * VARIABLES: G1 *

540   LET G1 = 0
550   LET W = W - 1
560   IF W <= 0 THEN 590
570   LET G1 = G1 + LOG(W)        Sum logs of factors
580   GO TO 550
590   IF W = 0 THEN 610
600   LET G1 = G1 + 0.57236494    Add ln Γ(½) = ln √π

610   REM   * OUTPUT: G1, LOG GAMMA (W) *
```

Algorithm 8.6 *Log Gamma Function*

The cumulative distribution functions may be evaluated using Simpson's rule, as was done in Algorithm 8.3 for the normal distribution. Serious difficulties arise in the integration for χ^2 with one degree of freedom and for F when $k_1 = 1$. In both these cases the integrand is singular at the origin, and the number of strips required in the integration is prohibitively large even to obtain very approximate results.

8.6.1   The *t* distribution

An exact algorithm (see Abramowitz and Stegun (1972)) for finding the probability p that corresponds to a given value of t with k degrees of freedom can be constructed from a series summation. If we define $\theta = \tan^{-1}(t/\sqrt{k})$, then $p = \frac{1}{2}(1 + A)$, where

$$A = 2\theta/\pi \qquad (k = 1),$$

$$A = \frac{2}{\pi}\left[\theta + \sin\theta\left(\cos\theta + \frac{2}{3}\cos^3\theta + \ldots + \frac{2.4 \ldots (k-3)}{1.3 \ldots (k-2)}\cos^{k-2}\theta\right)\right]$$

$$(k > 1 \text{ and odd}),$$

$$A = \sin\theta\left[1 + \frac{1}{2}\cos^2\theta + \frac{1.3}{2.4}\cos^4\theta + \ldots + \frac{1.3 \ldots (k-3)}{2.4 \ldots (k-2)}\cos^{k-2}\theta\right]$$

$$(k \text{ even}).$$

Algorithm 8.7, *t Distribution Fnc*, carries out this summation.

8.6.2   The χ^2 distribution

The algorithm that we give for the χ^2 distribution function is based on an algorithm by Lau (1980) for the cumulative distribution of the gamma function. The gamma

```
500   REM   ** T DISTRIBUTION FNC **
510   REM   * INPUT: ARGUMENT TØ *
520   REM   * K1 DEGREES OF FREEDOM *
530   REM   * VARIABLES: A1, CØ, I, J1, J2, K2, PØ, SØ, T, T1, W, W1 *

540   LET A1 = Ø.36338Ø23
550   LET W = ATN(TØ/SQR(K1))
560   LET SØ = SIN(W)
570   LET CØ = COS(W)
580   LET W1 = K1 - 2*INT(K1/2)              K1 odd or even?
590   IF W1 = Ø THEN 7ØØ

6ØØ   REM   K1 ODD
610   LET T1 = W
620   IF K1 = 1 THEN 84Ø                     Special case, K1 = 1
630   LET T = SØ*CØ
640   LET T1 = T1 + T
650   IF K1 = 3 THEN 84Ø                     Special case, K1 = 3
660   LET J1 = Ø                             Parameters
670   LET J2 = 1                             for summation
680   LET K2 = (K1 - 3)/2
690   GO TO 77Ø

7ØØ   REM   K1 EVEN
710   LET T1 = SØ
720   IF K1 = 2 THEN 85Ø                     Special case, K1 = 2
730   LET T = SØ
740   LET J1 = -1                            Parameters
750   LET J2 = Ø                             for summation
760   LET K2 = (K1 - 2)/2

77Ø   REM   SERIES SUMMATION
780      FOR I = 1 TO K2
790      LET J1 = J1 + 2
8ØØ      LET J2 = J2 + 2
810      LET T = T*CØ*CØ*J1/J2
820      LET T1 = T1 + T
830      NEXT I
840   LET T1 = T1*(1 - A1*W1)
850   LET PØ = Ø.5*(1 + T1)

860   REM   * OUTPUT: PØ, PROBABILITY X < TØ *
```

Algorithm 8.7 *t Distribution Fnc*

distribution with parameters α, β (usually denoted by $G(\alpha, \beta)$) has probability density function (Hogg and Craig, 1970):

$$\frac{1}{\Gamma(\alpha)\beta^{\alpha}} x^{\alpha-1} e^{-x/\beta} \qquad 0 < x < \infty.$$

The parameter β is called a **scale parameter** since a transformation $y = cx$ changes only the value of β in the density function; α is called a **shape parameter**. The χ^2 distribution with k degrees of freedom is the gamma distribution $G(k/2, 2)$, as may be seen by comparing the probability density functions. In Algorithm 8.8, *Chi-squared Distribution Fnc*, the χ^2 distribution input values are modified in lines 550 and 560 and pass into an algorithm for the log gamma function. The algorithm from line 570 onwards may be treated as one for the distribution function of $G(K1, 1)$, with input values W and K1.

```
500   REM   ** CHI-SQUARED DISTRIBUTION FNC **
510   REM   * INPUT: ARGUMENT X2 *
520   REM   * K1 DEGREES OF FREEDOM *
530   REM   * REQUIRES: LOG GAMMA FUNCTION (8.6) AT 1500 *
540   REM   * VARIABLES: A2, G1, G2, K, P0, T, W, W1 *

550   LET W1 = 0.5*X2
560   LET K1 = 0.5*K1

570   REM   CALCULATE LOG GAMMA (K1 + 1)
580   LET W = K1 + 1
590   GOSUB 1500

600   REM   SERIES SUMMATION
610   LET G2 = 0
620   LET A2 = EXP(K1*LOG(W1) - G1 - W1)
630   IF A2 = 0 THEN 720
640   LET T = 1
650   LET G2 = 1
660   LET K = K1
670   LET K = K + 1
680   LET T = T*W1/K
690   LET G2 = G2 + T
700   IF T/G2 > 1.0E-6 THEN 670
710   LET P0 = G2*A2

720   REM   * OUTPUT: P0, PROBABILITY X < X2 *
```

Algorithm 8.8 *Chi-squared Distribution Fnc*

8.6.3 The *F* distribution
We give the algorithm *F Distribution Fnc* (A.3) in compressed form in Appendix B, page 142. It is based on an algorithm for the incomplete beta function:

$$I_x(a, b) = \frac{\Gamma(a)\Gamma(b)}{\Gamma(a+b)} \int_0^x t^{a-1}(1-t)^{b-1}\,\mathrm{d}t,$$

given by Majumder and Bhattacharjee (1973); the underlying method is to sum a series of integrations by parts. This integral equals the value of the F distribution function, with k_1 and k_2 degrees of freedom and with argument F when $a = k_2/2, b = k_1/2$ and $x = k_2/(k_2 + k_1 F)$.

8.6.4 Approximations
For many purposes an approximation to the value of the distribution function may be adequate. We give below transformations of t, χ^2 and F that will produce, in each case, a

variable z which is approximately a standard normal variable (i.e. one with mean 0 and variance 1). Thus for a given value of t, say, we can derive the corresponding value of z and then use an algorithm to find $P\emptyset$, the value of the distribution function corresponding to z, and hence to the original t value.

Wallace (1959) proposed for t with k degrees of freedom the transformation

$$z = \frac{8k+1}{8k+3}\left[k\ln\left(1+\frac{t^2}{k}\right)\right]^{1/2}.$$

For χ^2 with k degrees of freedom, Wilson and Hilferty (1931) showed that $(\chi^2/k)^{1/3}$ is approximately normal with mean $\left(1-\frac{2}{9k}\right)$ and variance $\left(\frac{2}{9k}\right)$. Hence

$$z = \frac{(\chi^2/k)^{1/3}-(1-2/(9k))}{(2/(9k))^{1/2}}.$$

This approximation may be applied to derive one for F. The F distribution with k_1 and k_2 degrees of freedom, $F_{(k_1, k_2)}$, is defined as the ratio (apart from a factor) of two χ^2 variables (Section 9.6). Thus we may write

$$F = (\chi_1^2/k_1)/(\chi_2^2/k_2),$$

the chi-squared distributions χ_1^2 and χ_2^2 having respectively k_1 and k_2 degrees of freedom. Hence

$$F^{1/3} = (\chi_1^2/k_1)^{1/3}/(\chi_2^2/k_2)^{1/3}$$

is the ratio of two variables which are approximately normally distributed. There is a result due to Geary (1930) concerning the ratio $v = z_1/z_2$, where z_1 and z_2 are normal variables with means μ_1, μ_2 and variances σ_1^2, σ_2^2 respectively: he showed that

$$z = \frac{\mu_1-\mu_2 v}{(\sigma_1^2+\sigma_2^2 v^2)^{1/2}}$$

is approximately a standard normal variable. If we substitute $F^{1/3}$ for v, and the appropriate values for μ_1, μ_2, σ_1^2, σ_2^2, we obtain

$$z = \frac{\left(1-\frac{2}{9k_2}\right)F^{1/3}-\left(1-\frac{2}{9k_1}\right)}{\left(\frac{2}{9k_2}F^{2/3}+\frac{2}{9k_1}\right)^{1/2}}.$$

Paulson (1942) appears to have been the first person to propose this. Algorithm 8.9, *F Distribution Fnc – Approx*, makes use of this result. It uses *Normal Distribution Fnc – Approx* (Algorithm 8.4) as a subroutine. Algorithm 8.9 is valid only for values of $F \geq 1$. If $F < 1$, then F is replaced by $1/F$, k_1 and k_2 are interchanged, and the resulting probability is subtracted from 1. A correction to the z value is required in the algorithm when $k_2 \leq 3$ in order to improve accuracy for these small values of k_2. With this correction incorporated, Algorithm 8.9 gives reasonably satisfactory results; for example it calculates upper tail probabilities in the region of 0.05 with a relative accuracy of a few percent over the whole range of values of k_1, k_2.

```
5ØØ   REM   ** F DISTRIBUTION FNC - APPROX **
51Ø   REM   * INPUT: ARGUMENT F (>= 1) *
52Ø   REM   * DEGREES OF FREEDOM K1, K2 *
53Ø   REM   * REQUIRES: NORMAL DISTRIBUTION FNC (8.4) AT 15ØØ *
54Ø   REM   * VARIABLES: A1, A2, W, W1, W2, Z, PØ *

55Ø   LET A1 = 2/(9*K1)
56Ø   LET A2 = 2/(9*K2)
57Ø   LET W = F↑(1/3)
58Ø   LET W1 = W + A1 - W*A2 - 1
59Ø   LET W2 = A2*W*W + A1
6ØØ   LET Z = W1/SQR(W2)

61Ø   IF K2 > 3 THEN 64Ø
62Ø   LET Z = Z*(1 + Ø.Ø8*(Z↑4)/(K2↑3))

63Ø   REM   FIND PØ CORRESPONDING TO NORMAL Z
64Ø   GOSUB 15ØØ

65Ø   REM   * OUTPUT: PØ, PROBABILITY X < F *
```

Correction for small k_2

Algorithm 8.9 *F Distribution Fnc – Approx*

We can also use this algorithm to derive approximate values of the distribution functions of t and χ^2, using the following relations:

$$t^2_{(k)} = F_{(1,\,k)}$$

and $\chi^2_{(k)}/k = F_{(k,\,\chi)}$,

where k denotes degrees of freedom.

Thus to find the value of the distribution function of t corresponding to a given value T of t, enter T^2 into Algorithm 8.9 using 1 and k as degrees of freedom; we obtain from the algorithm an approximate value of the required probability. When dealing with χ^2 we cannot of course put $k_2 = \infty$, but using a very large number for k_2 will be satisfactory. As the reader will see, it is possible by using these relations to produce a relatively short algorithm to give probabilities for normal, t, χ^2 and F variables.

8.6.5 Inverse distribution functions based on approximations

Approximate values of the inverse distribution functions of the sampling distributions can be found conveniently using the approximations of Section 8.6.4. We may find, for example, the value of t which corresponds to the value 0.95 of the distribution function.

If t has k degrees of freedom, the approximation of Section 8.6.4 can be rewritten

$$t_{(k)} = [k(\exp(w^2/k) - 1)]^{1/2},$$

in which $w = z(8k + 3)/(8k + 1)$. Algorithm 8.10, *Inverse t – Approx*, uses this relation; it requires *Inverse Normal – Approx* (Algorithm 8.5) as a subroutine. If we input the probability PØ and the degrees of freedom K1, Algorithm 8.10 calculates first the corresponding standard normal variable Z, and then goes on to calculate TØ using the relation above.

```
500   REM   ** INVERSE T - APPROX **
510   REM   * INPUT: PROBABILITY PØ, DEGREES OF FREEDOM K1 *
520   REM   * REQUIRES: INVERSE NORMAL (8.5) AT 1500 *
530   REM   * VARIABLES: TØ, W, Z *

540   REM   FIND NORMAL Z CORRESPONDING TO PØ
550   GOSUB 1500

560   LET W = Z*(1 + 2/(1 + 8*K1))
570   LET TØ = K1*(EXP(W*W/K1) - 1)
580   LET TØ = SQR(TØ)

590   REM   * OUTPUT: T-VALUE, TØ *
```

Algorithm 8.10 *Inverse t – Approx*

The *t* values derived from Algorithm 8.10 with input probabilities PØ of .975 and .995, which correspond to two-tail *t* tests at 5 % and 1 % significance levels, are compared with the exact values in Table 8.1. The algorithm is not reliable for fewer than four degrees of freedom.

Table 8.1

Degrees of freedom		1	2	5	10	20	30	60	120	∞
PØ = .975	calculated	17.6	4.48	2.58	2.23	2.09	2.04	2.00	1.98	1.96
	exact	12.7	4.30	2.57	2.23	2.09	2.04	2.00	1.98	1.96
PØ = .995	calculated	142.9	11.17	4.07	3.18	2.85	2.75	2.66	2.62	2.58
	exact	63.6	9.92	4.03	3.17	2.84	2.75	2.66	2.62	2.58

In a similar way, we may obtain χ_k^2 in terms of *z*:

$$\chi_k^2 = k\left(1 - \frac{2}{9k} + z\sqrt{\frac{2}{9k}}\right)^3.$$

Algorithm 8.11, *Inverse Chi-squared – Approx*, uses this equation. Calculated and exact χ^2 values corresponding to probabilities PØ of .95 and .99 are given in Table 8.2.

```
500   REM   ** INVERSE CHI-SQUARED - APPROX **
510   REM   * INPUT: PROBABILITY PØ, DEGREES OF FREEDOM K1 *
520   REM   * REQUIRES: INVERSE NORMAL (8.5) AT 1500 *
530   REM   * VARIABLES: A1, W, X2, Z *

540   REM   FIND NORMAL Z CORRESPONDING TO PØ
550   GOSUB 1500

560   LET A1 = 2/(9*K1)
570   LET W = 1 - A1 + Z*SQR(A1)
580   LET X2 = K1*(W↑3)

590   REM   * OUTPUT: CHI-SQUARED VALUE X2 *
```

Algorithm 8.11 *Inverse Chi-squared – Approx*

Table 8.2

Degrees of freedom		1	2	5	10	20	30	60
$P\emptyset = .95$	calculated	3.75	5.94	11.04	18.29	31.4	43.8	79.1
	exact	3.84	5.99	11.07	18.31	31.4	43.8	79.1
$P\emptyset = .99$	calculated	6.59	9.23	15.13	23.2	37.6	50.9	88.4
	exact	6.63	9.21	15.09	23.2	37.6	50.9	88.4

The relation between z and F in Section 8.6.4 yields a quadratic in $F^{1/3}$, when a particular value is substituted for z. One root is positive and the other negative; the negative root is rejected. This calculation is carried out in Algorithm 8.12, *Inverse*

```
500   REM   ** INVERSE F - APPROX **
510   REM   * INPUT: PROBABILITY PØ (>= Ø.5) *
520   REM   * DEGREES OF FREEDOM K1, K2 *
530   REM   * REQUIRES: INVERSE NORMAL (8.5) AT 15ØØ *
540   REM   * VARIABLES: A1, A2, F, W, W1, W2, W3, W4, Z *

550   REM   FIND NORMAL Z CORRESPONDING TO PØ
560   GOSUB 15ØØ

570   LET A1 = 2/(9*K1)
580   LET A2 = 2/(9*K2)
590   LET W = Z*Z

600   LET W1 = 1 + A2*(A2 - W - 2)          Coefficients of
610   LET W2 = A1 + A2 - A1*A2 - 1          quadratic
620   LET W3 = 1 + A1*(A1 - W - 2)
630   LET W4 = SQR(W2*W2 - W1*W3)
640   LET F = (W4 - W2)/W1                  Positive root
650   LET F = F↑3

660   REM   * OUTPUT: F-VALUE, F *
```

Algorithm 8.12 *Inverse F – Approx*

F – Approx. Table 8.3 gives calculated and exact F values corresponding to probabilities $P\emptyset$ of .95 and .99. The algorithm is not reliable when k_2 is less than four. Ashby (1968) proposed improving the value F given by the approximation, for $k_2 \leq 10$, by using a linear function $mF + c$ where m and c are constants depending on the significance level and on k_2 (but not on k_1). He gives a table of values of m and c for probabilities of .95, .99 and .999. No easy functional representation of the values appears to be possible so they have to be stored as a table in the computer. A substantial improvement in Algorithm 8.12, for $k_2 \leq 3$, can be obtained by replacing the value of z obtained from the normal subroutine with

$$z' = k_2^{3/4} u \, (1.1581 - 0.2296u - 0.0042u^2 - 0.0027u^3)$$

where $u = z/k_2^{3/4}$.

Table 8.3 Comparison of *F* values calculated by *Inverse F−Approx* (Algorithm 8.12) with exact values. The *F* distribution has k_1 and k_2 degrees of freedom; P∅ is the probability (percentage point) for which the *F* value is required.

P∅ = .95

k_2	k_1	1	2	5	10	30	60
2	calculated	17.20	18.25	18.83	19.01	19.13	19.16
	exact	18.51	19.00	19.30	19.40	19.46	19.48
5	calculated	6.56	5.82	5.11	4.80	4.56	4.50
	exact	6.61	5.79	5.05	4.74	4.50	4.43
30	calculated	4.07	3.29	2.53	2.16	1.84	1.74
	exact	4.17	3.32	2.53	2.16	1.84	1.74

P∅ = .99

k_2	k_1	1	2	5	10	30	60
2	calculated	71.31	78.85	83.97	85.79	87.04	87.36
	exact	98.50	99.00	99.30	99.40	99.47	99.48
5	calculated	16.76	14.01	11.81	10.93	10.29	10.13
	exact	16.26	13.27	10.97	10.05	9.38	9.20
30	calculated	7.50	5.39	3.71	2.98	2.39	2.21
	exact	7.56	5.39	3.70	2.98	2.39	2.21

8.6.6 Inverse *F* distribution

To calculate the exact value of *F* corresponding to a particular probability requires the solution of the equation

$$F(x) = p_0,$$

where $F(x)$ is the distribution function of *F* and p_0 is the chosen probability. The solution may be obtained using Newton iteration; we give an appropriate algorithm, *Inverse F* (A.4), in Appendix B. It uses the *F* distribution function derived from the incomplete beta function (Algorithm A.3).

8.7   Test data for Algorithms

Algorithm 8.1, *Binomial Probabilities*. Input: N = 3, P = 0.4. Output: P(·) = (.216, .432, .288, .064).

Algorithm 8.2, *Simpson's Rule*. Input: Define a function $0.5x^2 - 5x + 14$, L∅ = 2, U∅ = 14, N∅ = 5. Output: P∅ = 144.

Algorithm 8.3, *Normal Distribution Fnc − Integration*. Input: (i) Z = 1, N∅ = 5. (ii) Z = 2, N∅ = 11. Output: P∅ = (i) 0.841; (ii) 0.977.

Algorithm 8.4, *Normal Distribution Fnc − Approx*. Input: Z = (i) 1, (ii) 2. Output: P∅ = (i) 0.841, (ii) 0.977. Compare with output for test of Algorithm 8.3.

Algorithm 8.5, *Inverse Normal − Approx*. Input: P∅ = (i) 0.05; (ii) 0.95; (iii) 0.99.

Output: Z = (i) − 1.64449 (exact − 1.64485); (ii) 1.64449; (iii) 2.32765 (exact 2.32635).
Algorithm 8.6, *Log Gamma Function.* Input: W = (i) 4, (ii) 5.5. Output:
G1 = (i) 1.79176, (ii) 3.95781.
Algorithm 8.7, *t Distribution Fnc.* Input: (i) T0 = 2.776, K1 = 4. (ii) T0 = 2.528,
K1 = 20. Output: P0 = (i) 0.974989; (ii) 0.990000.
Algorithm 8.8, *Chi-squared Distribution Fnc.* Input: (i) X2 = 13.28, K1 = 4.
(ii) X2 = 31.41, K1 = 20. Output: P0 = (i) 0.990014, (ii) 0.949994.
Algorithm 8.9, *F Distribution Fnc – Approx.* Input: (i) F = 18.51, K1 = 1, K2 = 2.
(ii) F = 4.41, K1 = 20, K2 = 10. Output: P0 = (i) 0.952772 (exact .95); (ii) 0.989502
(exact .99).
Algorithm 8.10, *Inverse t – Approx.* Input: (i) P0 = 0.95, K1 = 10. (ii) P0 = 0.99,
K1 = 20. Output: T = (i) 1.812 (exact 1.812); (ii) 2.530 (exact 2.528).
Algorithm 8.11, *Inverse Chi-squared Approx.* Input: (i) P0 = 0.95, K1 = 10.
(ii) P0 = 0.99, K1 = 20. Output: X2 = (i) 18.2894 (exact 18.3070); (ii) 37.6040 (exact
37.5662).
Algorithm 8.12, *Inverse F – Approx.* Input: (i) P0 = 0.95, K1 = 10, K2 = 15.
(ii) P0 = 0.99, K1 = 15, K2 = 10. Output: F = (i) 2.54 (exact 2.54); (ii) 4.62
(exact 4.56).

8.8   Exercises

1   Write a program, based on Algorithm 8.1, *Binomial Probabilities*, to calculate the
probability that a binomial random variable is less than or equal to X given the values of
X, N and P (the cumulative distribution function).

2   Write a program to calculate the probability mass function of the Poisson
distribution, as described in Section 8.2.2. Run the program and generate the
probabilities, up to P(3), for a distribution with parameter $\lambda = 2/3$. (The values are
P(0) = 0.5134, P(1) = 0.3423, P(2) = 0.1141, P(3) = 0.0254; the total probability of
higher values is 0.0048.)

3   Write a program to calculate the probability mass function of the geometric
distribution, as described in Section 8.2.3. Run the program and generate the
probabilities, up to P(12), for a distribution with parameter $p = 1/4$. (The values are
0.25, 0.1875, 0.1406, 0.1055, 0.0791, 0.0593, 0.0445, 0.0334, 0.025, 0.0188, 0.0141, 0.0106;
the total probability of higher values is 0.0316.)

4   Use the functions defined in Section 8.3 to find the probability density function and
the cumulative distribution function for the exponential distribution with parameter
$\lambda = 1.2$ for $x = 0.25, 0.5, 0.75, 1.0, 1.25, 1.5, 2.0, 3.0, 5.0$.

5   Use the function defined at the end of Section 8.3 to find the median, and upper and
lower quartiles, of the exponential distribution with parameter $\lambda = 0.8$.

6   The binomial probabilities arising from a distribution in which n is large and p is
small can be approximated by the corresponding Poisson probabilities, with $\lambda = np$.
Write a program to calculate the probabilities up to P(10), by both methods, and the
differences between the values found using each of the two methods. Print the results in

a convenient table to be used for comparison of the exact and approximate probabilities.

Run the program for (i) $n = 100$, $p = 0.01$; (ii) $n = 50$, $p = 0.04$; (iii) $n = 25$, $p = 0.05$.

7 Write a program which uses Algorithm 8.2, *Simpson's Rule*, to find the (cumulative) distribution function of the random variable X whose probability density is $f(x)$ $= \dfrac{1}{108}x(6-x)^2$ for $0 \le x \le 6$ and zero elsewhere. Print out values of the distribution function at intervals of 0.5 for X.

The **mean** of a continuous distribution is given by $\int x f(x)\, dx$. Write a program to find this in a similar way. (The exact value is 2.4.)

8* Write a program which uses Algorithm 8.2, *Simpson's Rule*, to find the distribution function of the *t*-distribution. (Compare with Algorithm 8.3.)

9 If a variable Y follows a normal distribution with mean μ and variance σ^2 (i.e. standard deviation σ) then $Z = (Y-\mu)/\sigma$ follows the standard normal distribution, with mean 0 and variance 1 (*ABC*, Section 14.5); Z is the normal variable for which algorithms have been given in Section 8.5. Modify Algorithms 8.3 (or 8.4) and 8.5 to deal with a variable Y whose mean and variance are given as part of the input.

Run these programs to answer the following questions.
(a) If Y has mean 100 and standard deviation 15, find (i) $P(Y > 120)$, (ii) $P(Y < 125)$, (iii) $P(85 < Y < 100)$, (iv) $P(110 < Y < 140)$, (v) $P(65 < Y < 75)$, (vi) $P(Y > 90)$.
(b) If Y has mean 5 and variance 4, find the deciles of the distribution of Y (see Section 3.7).

10 Use Algorithm 8.10, *Inverse t – Approx*, to generate a *t*-table giving the values of *t* corresponding to P∅ = 0.95, 0.975, 0.99, 0.995, 0.999 and 0.9995, and print the table to show these in six columns; the rows of the table should correspond to the degrees of freedom. Label the table so that it could be used in the same way as a *t*-table printed in a book (e.g. *ABC*, Table A3).

11 Use Algorithm 8.11, *Inverse Chi-squared – Approx*, to generate a χ^2 table giving the values of χ^2 corresponding to P∅ = 0.995, 0.975, 0.05, 0.025, 0.01, 0.005, 0.001, and print the table to show these in seven columns; the rows of the table should correspond to the degrees of freedom. Label the table so that it could be used in the same way as a χ^2-table printed in a book (e.g. *ABC*, Table A4).

9 Generation of random samples from non-uniform distributions

9.1 Introduction

When carrying out stochastic simulations, we need to be able to draw random samples from a wide range of statistical distributions. For example, if we wish to simulate people arriving at random at a railway station ticket office, we might make the time interval between arrivals an exponential random variable. Likewise, if we are simulating the engineering performance of a temporary footbridge, we might assume that the weights of the people crossing it are normally distributed. We shall see that the basis of all the methods of sampling from distributions is a source of random samples from the uniform distribution U(0, 1), as described in Chapter 7, and we shall therefore assume that these are available from an RND function on the computer.

9.2 Simulation of the behaviour of a queue

Suppose we wish to discover the effect on the queue at a railway booking office of allowing customers to pay for their tickets by cheque as an alternative to paying by cash. To program a simple queueing system such as this we need to specify, or be able to calculate, the arrival time A_i and the service time S_i of each customer i. Let us call the time at which the service of the ith customer is completed the 'leaving time', L_i. At this instant, either there is a queue, in which case the leaving time of the $(i + 1)$th customer is $L_i + S_{i+1}$, or there is no queue, in which case the leaving time of the $(i + 1)$th customer is $A_{i+1} + S_{i+1}$. These calculations form the core of the simulation and are programmed in lines 280–310 of Program 9.1, *Booking Office Simulation*.

In our model we shall assume that the time interval between successive customers arriving follows an exponential distribution, which is appropriate if they are arriving at random instants in time but at a constant overall rate. The random sample from the distribution is chosen in line 190; the expression used there is justified in Section 9.3. The service time S_i for a customer will depend on whether he pays by cheque or by cash. We shall make the very simple assumption that the service time by each of these methods is a fixed value (the value for cheque obviously being greater than that for cash); it would, of course, in practice be more realistic to assume that each of these two service times had a distribution. The probability that any individual customer will pay by cheque is fixed at the beginning of the simulation, and the random number function RND is used to decide whether a particular customer pays by cheque or cash; this determines his service time (see lines 220–260 of Program 9.1).

The variables W and S are introduced to record interesting characteristics of each run of the simulation: W records the sum of the times that individual customers spend waiting in the queue, while S is the sum of the service times of the customers. The calculations are carried out in lines 330–350; and the mean values, obtained by dividing the sums W and S by the total number of customers N, are output at the end of the

```
10   REM  ** BOOKING OFFICE SIMULATION **
20   REM  * VARIABLES: A, A1, C, L, N, N1, N2 *
30   REM  * S, S1, T, T1, T2, T3, W, W1 *
40   DEF FNR(X) = INT(X*10 + 0.5)/10

50   INPUT "SERVICE TIME - CASH? "; T1
60   INPUT "SERVICE TIME - CHEQUE? "; T2
70   INPUT "PROBABILITY OF CHEQUE? "; C
80   INPUT "MEAN INTER-ARRIVAL TIME? "; T3
90   INPUT "DURATION OF RUN? "; T

100  LET N = 0                              Number of customers
110  LET N1 = 0                             Number of cash customers
120  LET N2 = 0                             Number of cheque customers
130  LET A = 0                              Arrival time
140  LET L = 0                              Leaving time
150  LET S = 0                              Total service time
160  LET W = 0                              Total waiting time

170  LET N = N + 1
180  REM  CALCULATE INTER-ARRIVAL TIME A1
190  LET A1 = -T3*LOG(RND(1))
200  LET A = A + A1                         Arrival time
210  REM  CALCULATE SERVICE TIME S1
220  IF RND(1) < C THEN 260                 Decide cash or cheque
230  LET S1 = T1                            Cash service time
240  LET N1 = N1 + 1                        Count cash customers
250  GO TO 280
260  LET S1 = T2                            Cheque service time
270  LET N2 = N2 + 1                        Count cheque customers

280  IF A > L THEN 310                      Test whether queue formed
290  LET L = L + S1                         Leaving time, if queue
300  GO TO 320
310  LET L = A + S1                         Leaving time, if no queue

320  LET W1 = L - A - S1                    Waiting time of a customer
330  LET W = W + W1                         Total waiting time
340  LET S = S + S1                         Total service time

350  IF A < T THEN 170                      Test whether run finished

360  PRINT
370  PRINT "NUMBER OF CUSTOMERS "; N
380  PRINT "CASH CUSTOMERS        "; N1
390  PRINT "CHEQUE CUSTOMERS      "; N2
400  PRINT "MEAN WAITING TIME     "; FNR(W/N)
410  PRINT "MEAN SERVING TIME     "; FNR(S/N)
420  PRINT "TOTAL IDLE TIME       "; FNR(L - S)

430  STOP
```

Program 9.1 *Booking Office Simulation*

program. The numbers of cash and cheque customers are recorded in N1 and N2 respectively. The total time for which the server (i.e. the clerk in the booking office) is not actually serving is also of interest: it is output as 'total idle time'. Means are rounded to 1 decimal place on output, using the rounding function defined in line 40.

Let us consider a particular run of this simulation. Suppose that we wish to simulate what happens at a booking office between 8.00 and 9.00 a.m. Using seconds as units, the duration of this run is 3600. Reasonable values for service times are, for cash, 15 seconds and, for cheque, 45 seconds. Let us take the mean time-interval between arrivals as 20 seconds. Five runs of this simulation were made with no customers paying by cheque, and the mean waiting times were

> 20.5, 18.5, 26.4, 19.9, 17.1 seconds.

Five more runs were made, under the same conditions except that there was a probability 0.10 that any individual customer would offer a cheque. The mean waiting times now were

> 178.0, 72.7, 79.3, 35.0, 55.5 seconds.

The magnitude of the mean waiting time is much greater in the second set of runs than in the first, and so is the variability in waiting time.

It would be interesting to look at the distribution of the waiting times of individual customers within a run, and this information would be easy to obtain. If we wish to make a number of runs with the same parameter values, we can add an input statement at the end of the program asking whether the program should be run again; if so, a jump to line 100 will give another run with the same parameter values.

Program 9.1, with changes of detail, would make a general program for any single-server queue simulation. In a general program it would probably be wise to use subroutines to calculate the inter-arrival times (i.e. the time-interval between successive customers arriving) and the service times, so that the distributions used could be changed easily.

9.3 A general method of sampling statistical distributions

The simplest random variable to sample by computer, given a supply of uniformly distributed random variables, is the one that takes the value 1 with probability p and the value 0 with probability $(1 - p)$. This is known as the Bernoulli distribution (*ABC*, Section 7.5); it may be regarded as a binomial distribution in which the parameter n equals 1. Bernoulli distributions might be used in a simulation of a game of tennis (*ABC*, Section 11.6). Thus we might fix the probability that one of the players wins directly from a serve at 0.1. Then, in the simulation, we sample a Bernoulli distribution with $p = 0.1$; if the value of the random variable is 1, the player chosen wins directly from his serve, but does not if the value is zero. Similarly, we might fix the probability that a particular player wins a rally and again decide the progress of a game by sampling a Bernoulli distribution. To choose the sample from the Bernoulli distribution, given a uniform variable u, we make the variable 1 if $u < p$ and 0 otherwise.

It may help the reader to understand better the process of drawing a random sample from a statistical distribution if we consider how it was sometimes done before

computers were common. To draw a random sample from a Bernoulli distribution with
$p = 0.1$ we might put ten discs in a hat, label one of them '1' and label the remaining nine
'0'. After a thorough shake of the hat, the label of '0' or '1' on a disc chosen from the hat
would give us the sample value. Let us now consider sampling a more complex
distribution, a binomial distribution with $n = 3$, $p = 0.4$. The probabilities of obtaining
each possible value of the variable are given in the second line of Table 9.1, and the
cumulative probabilities in the third line.

Table 9.1

x	0	1	2	3
$\Pr(X = x)$	.216	.432	.288	.064
$\Pr(X \leq x)$	.216	.648	.936	1.000

We might draw the sample by putting into a (very large) hat 1000 discs, of which 216
are marked '0', 432 marked '1', 288 marked '2' and 64 marked '3'. It would not be
necessary to have special marks on the discs provided that they were numbered 1–1000.
If we draw out of the hat any of the discs numbered 1–216, then we use a sample value of
0; the discs 217–648 give a sample value of 1; 649–936 give 2; and 937–1000 give 3.

We can represent this method using a graph (Figure 9.1). Choosing a disc corresponds to choosing a number (413 in Figure 9.1) between 1 and 1000, and may be

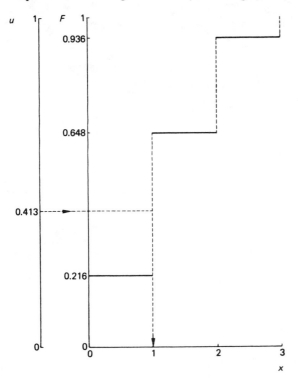

Figure 9.1

denoted by a point on the left-hand axis. The fact that 413 falls in the interval 217–648 inclusive, and therefore corresponds to the sample value 1, is seen on the graph by drawing a line, parallel to the x-axis, to meet the graph. Now the graph is, in fact, the graph of $F(x)$, the cumulative distribution function of X; and we see that if the random number u, when scaled to lie between 0 and 1, falls in the interval

$$F(x_{i-1}) < u \leqslant F(x_i),$$

then x_i is the appropriate sample value from the distribution. This is a general method that may be used with any discrete distribution.

By considering a similar graph, we may develop a method of choosing a random sample from a continuous variable. Imagine a discrete random variable in which the intervals between successive values are very small. The graph of $F(x)$ then changes by steps, but very small ones, and in the limit the graph, for a continuous distribution function, will be a continuous line (Figure 9.2). As with the discrete random variable, we may proceed in a similar way to pass from the uniform random variable u (with a value of .237 in the example) to the value x (.270 in the example) of the random sample. The required sample value x comes from solving the equation

$$u = F(x),$$

where $F(x)$ is the cumulative distribution function of the random variable. The solution

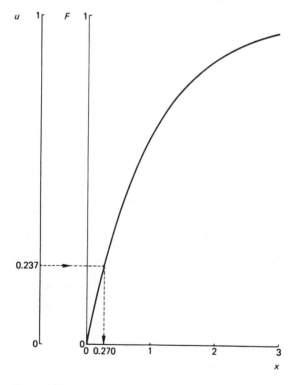

Figure 9.2

may be written in terms of the inverse distribution function as

$$x = F^{-1}(u).$$

For example, the exponential function has the cumulative distribution function (Section 8.3)

$$F(x) = 1 - \exp(-\lambda x).$$

To obtain a random value x from the distribution, given a uniform random variable u, we solve

$$u = 1 - \exp(-\lambda x),$$

to obtain

$$x = -\ln(1-u)/\lambda.$$

Since u is $U(0, 1)$ then $(1-u)$ is also $U(0, 1)$. We may therefore make the argument of the log function the uniform random variable we have generated. Hence a random sample from the exponential distribution, with parameter denoted by L, is obtained by the single line of programming:

```
10   X = -LOG(U)/L
```

The mean of this exponential distribution will be $1/L$ (*ABC*, page 183). The method we have given is a general one for continuous random variables provided that we can obtain an explicit expression for the inverse distribution function $F^{-1}(x)$.

We have begun with the general methods. However, methods based on the particular properties of distributions are often more convenient and efficient. We shall therefore look individually at the statistical distributions that are commonly used in simulation.

9.4 Discrete random variables

9.4.1 Bernoulli distribution

This serves as a model for any situation in which we can think of a trial being made with probability p of 'success' and probability $(1-p)$ of 'failure' (*ABC*, Section 7.5). Thus we may use it to model such widely different situations as the toss of a coin ('success' being a Head), or whether a surgical operation is successful or not. The standard definition gives the variable X a value of 1 with probability p and a value 0 with probability $(1-p)$, and we follow this definition in Algorithm 9.2, *Random – Bernoulli*.

```
500  REM  ** RANDOM - BERNOULLI **
510  REM  * INPUT: PARAMETER P *
520  REM  * VARIABLES: X *

530  IF RND(1) > P THEN 560
540  LET X = 1
550  GO TO 570
560  LET X = 0

570  REM  * OUTPUT: BERNOULLI VARIABLE X *
```

Algorithm 9.2 *Random – Bernoulli*

9.4.2 Binomial distribution

The probability mass function of this variable has already been given in Section 8.2 (see also *ABC*, Section 7.6). The variable may be used to model, for example, the number of defective components in random samples of 10 taken from an industrial production line, or the number of blue-eyed players (as opposed to non-blue-eyed players) in cricket teams. The most convenient way to generate random samples is to use the property that a binomial distribution with parameters n and p is the sum of n Bernoulli variables, each with parameter p. Thus we take a $U(0, 1)$ variable u. If $0 < u < p$ count 1, otherwise count 0. Repeat n times, and the sum of the counts is a binomial variable. This is programmed in Algorithm 9.3, *Random – Binomial*.

```
500  REM  ** RANDOM - BINOMIAL **
510  REM  * INPUT: PARAMETERS N1, P *
520  REM  * VARIABLES: I, X *

530  LET X = 0
540      FOR I = 1 TO N1
550      IF RND(1) > P THEN 570
560      LET X = X + 1
570      NEXT I

580  REM  * OUTPUT: BINOMIAL VARIABLE X *
```

Algorithm 9.3 *Random – Binomial*

9.4.3 Poisson distribution

This may be used, for example, to model the number of misprints on the page of a book, or the number of cars that pass an observation point on a road in one minute. It is described in *ABC*, Chapter 19; the probability mass function has already been given on page 74 of this book.

We use the fact that if events occur at random in time, then the interval between events has an exponential distribution and the number of events in a fixed interval of time has a Poisson distribution. If we have a source of random variables from an exponential distribution, which are easy to generate, and we find how many just fit into a unit time interval, then that number will have a Poisson distribution. More precisely, if an exponential variable is labelled v_i, and if we fix X such that

$$\sum_{i=0}^{X-1} v_i < 1 \le \sum_{i=0}^{X} v_i \qquad (X = 0, 1, \ldots)$$

then X has a Poisson distribution. This Poisson distribution will have parameter λ (and therefore mean λ) if we sample from an exponential distribution with parameter λ, and therefore mean $1/\lambda$.

To obtain a random variable v_i from an exponential distribution with parameter λ, using a uniform random variable u_i, we put (see page 95)

$$v_i = -(\ln u_i)/\lambda.$$

Thus the Poisson variable X we require is given by

$$\sum_{i=1}^{X} -(\ln u_i)/\lambda < 1 \le \sum_{i=1}^{X+1} -(\ln u_i)/\lambda$$

or $\quad \ln \prod_{i=0}^{X-1} u_i > -\lambda \ge \ln \prod_{i=0}^{X} u_i$

or $\quad \prod_{i=0}^{X-1} u_i > e^{-\lambda} \ge \prod_{i=0}^{X} u_i.$

We take a uniform variable u_0; if it is less than or equal to $e^{-\lambda}$ we stop, and take $X = 0$. If $u_0 > e^{-\lambda}$, we form the product $u_0 u_1$; if this is less than or equal to $e^{-\lambda}$, we take $X = 1$. Otherwise we continue until $u_0 u_1 u_2 \ldots u_x$ is the first product that is less than or equal to $e^{-\lambda}$, which leads us to take $X = x$. This method is programmed in Algorithm 9.4, *Random – Poisson*.

```
500   REM   ** RANDOM - POISSON **
510   REM   * INPUT: PARAMETER L *
520   REM   * VARIABLES: P∅, W, X *

530   LET X = ∅
540   LET W = 1
550   LET P∅ = EXP(-L)

560   LET W = W*RND(1)
570   IF W <= P∅ THEN 600
580   LET X = X + 1
590   GO TO 560

600   REM   * OUTPUT: POISSON VARIABLE X *
```

Algorithm 9.4 *Random – Poisson*

9.4.4   Geometric distribution
If independent Bernoulli trials are made until a 'success' occurs, then the total number of trials required is a geometric random variable. The distribution is described in *ABC*, Section 7.7, and its probability mass function has already been given on page 75 of this book. It could serve, for example, as a model for the number of cars passing an

```
500   REM   ** RANDOM - GEOMETRIC **
510   REM   * INPUT: PARAMETER P *
520   REM   * VARIABLES: X *

530   LET X = 1
540   IF RND(1) <= P THEN 570
550   LET X = X + 1
560   GO TO 540

570   REM   * OUTPUT: GEOMETRIC VARIABLE X *
```

Algorithm 9.5 *Random – Geometric*

observation point up to the first car of a particular make: record this number and begin counting again. Algorithm 9.5, *Random – Geometric*, is based directly on this definition. If P is small this algorithm will be slow; Knuth (1981) suggests an alternative. Algorithm 9.5 is replaced by an algorithm whose operative statements are

```
10   W = RND(1)
20   X = INT(LOG(W)/LOG(1-P)) + 1
```

9.5   Normal distribution

The normal distribution is the most important distribution in statistics. It is discussed in *ABC*, Chapter 14; the probability density function has already been given on page 77 of this book. It is likely to provide a good model for a variate when:

1) there is a strong tendency for the variate to take a central value;
2) positive and negative deviations from this central value are equally likely;
3) the frequency of deviations falls off rapidly as the deviations become larger.

It also provides a good approximation to many other statistical distributions over certain ranges of their parameters. Examples of variates which it might model are the weights of packets of cornflakes and the heights of members of a football team.

We confine ourselves to obtaining random samples from the standard normal distribution, i.e. the distribution with mean 0 and variance 1 whose values are often represented by z. If we require random samples from the normal variable x which has mean μ and variance σ^2, we derive x from z by the relation

$$x = \mu + z\sigma.$$

9.5.1   Use of the central limit theorem

The simplest method of generating normal variables is an approximate one, and relies on the central limit theorem (*ABC*, Section 15.1). A U(0, 1) variable, which we denote by u, has mean 0.5 and variance $\frac{1}{12}$ (*ABC*, Section 13.5). Hence $\sum_{i=1}^{n} u_i$, the sum of n independent uniform variables, has mean $n/2$ and variance $n/12$ (*ABC*, Section 15.2). Moreover, by the central limit theorem, this sum is approximately normally distributed

```
500   REM   ** RANDOM - NORMAL (CLT) **
510   REM   * VARIABLES: I, T, X *

520   LET T = 0
530       FOR I = 1 TO 12
540       LET T = T + RND(1)
550       NEXT I
560   LET X = T - 6

570   REM   * OUTPUT: STANDARD NORMAL VARIABLE X *
```

Algorithm 9.6 *Random – Normal (CLT)*

provided that n is sufficiently large; in this particular case, the value 12 of n is adequate and has the convenience of keeping the arithmetic simple. Thus the sum of 12 independent uniform variables is approximately normal with mean 6 and variance 1. Algorithm 9.6, *Random − Normal (CLT)*, uses this result to produce a standard normal variable (with mean 0 and variance 1). The algorithm cannot produce values outside the range $(-6, 6)$, but this is not likely to matter since such values have an extremely low probability of occurring in the standard normal distribution.

9.5.2 The Box–Muller method

The second method we give is exact and was proposed by Box and Muller. Two independent uniform variables U(0, 1) are used to generate two independent standard normal variables N(0, 1). The method is based on the fact that if u_1, u_2 are independent samples from U(0, 1), then

$$z_1 = (-2 \ln u_1)^{1/2} \sin 2\pi u_2,$$
$$z_2 = (-2 \ln u_1)^{1/2} \cos 2\pi u_2$$

are independent samples from N(0, 1). (A proof is given in Fishman, 1978.) The method is programmed in Algorithm 9.7, *Random − Normal (Box–Muller)*. The method is exact only if the two uniform variables are independent; if the random numbers are generated

```
500   REM  ** RANDOM - NORMAL (BOX-MULLER) **
510   REM  * VARIABLES: A1, W1, W2, X1, X2 *

520   LET A1 = 6.283185308
530   LET W1 = RND(1)
540   LET W2 = RND(1)

550   LET W1 = -2*LOG(W1)
560   LET W1 = SQR(W1)
570   LET W2 = A1*W2
580   LET X1 = W1*SIN(W2)
590   LET X2 = W1*COS(W2)

600   REM  * OUTPUT: STANDARD NORMAL VARIABLES X1, X2 *
```

Algorithm 9.7 *Random − Normal (Box−Muller)*

by a linear congruential generator, for example, this will not be strictly true. For most purposes the results will be satisfactory and therefore we have used successive values of an RND function in our algorithm. The safest way to resolve the difficulty is to use two independent random number generators and take u_1 from one and u_2 from the other. Another approach, using a single random number generator, is not to take successive values from the generator but to allow a gap of up to (say) five, the exact size chosen at random, between one number and the next.

9.6 Distributions based on the normal distribution

Random variables from the sampling distributions χ^2, t and F may easily be generated using the definition of these variables in terms of normal variables. Using z_i to denote a standard normal variable then, provided that all the z_i's are independent,

1) the chi-squared variable with k degrees of freedom is given by

$$\chi^2 = z_1^2 + z_2^2 + \ldots + z_k^2;$$

2) the t variable with k degrees of freedom is given by

$$t = z_0 / \sqrt{(z_1^2 + z_2^2 + \ldots + z_k^2)};$$

3) the F distribution with k_1 and k_2 degrees of freedom is given by

$$F = \frac{\chi_{k_1}^2 / k_1}{\chi_{k_2}^2 / k_2}.$$

The **lognormal distribution** is sometimes useful in simulations. It has been used, for example, as a model for the distribution of (a) personal incomes, (b) age at first marriage, (c) tolerance to poison in animals. The lognormal variable Y is defined as a variable such that $\ln Y$ has a normal distribution. Thus if x is a sample from a normal distribution, then

$$y = e^x$$

is a sample from a lognormal variable.

9.7   Other continuous distributions

9.7.1   Exponential distribution

The generation of random samples from the exponential distribution was discussed in Section 7.3. The method is repeated here for reference. The distribution is described in *ABC*, Section 13.6; the probability density function is given on page 75 of this book. The distribution is frequently used to model the time interval between successive random events. Given a sample u from a uniform variable, a sample from an exponential distribution with parameter λ is given by

$$x = -(\ln u)/\lambda.$$

Since the mean, μ, of the exponential distribution is equal to $1/\lambda$, we can also write the random sample in the form $x = -\mu \ln u$.

9.7.2   Gamma distribution

The density function of the exponential distribution has a mode at zero. When modelling the distribution of the life-times of a product such as an electric light bulb (i.e. the length of time the bulb lasts before failing), or the serving-time taken at a ticket office or a shop counter, a distribution having a mode at a greater value than zero is useful; the gamma distribution is one such. The probability density function $G(\alpha, \beta)$ was given on page 81. We restrict ourselves to gamma distributions having an integer value for the 'shape' parameter α.

We use the properties that the sum of k independent variables, each from a $G(\alpha, \beta)$ distribution, is a $G(k\alpha, \beta)$ distribution (Hogg and Craig, 1970); and that $G(1, \beta)$ is an exponential distribution with parameter $1/\beta$ (that is with mean β). Thus, denoting uniform random variables by u_i,

$$x = -\beta \sum_{i=1}^{\alpha} \ln u_i = -\beta \ln \prod_{i=1}^{\alpha} u_i$$

is a sample from $G(\alpha, \beta)$.

Note that a χ^2 distribution with k degrees of freedom is $G\left(\dfrac{k}{2}, 2\right)$.

[*Note*: We give no test data for the algorithms of this chapter, but some of the exercises below will serve this purpose.]

9.8 Exercises

1 Write a program to simulate the progress of a game of tennis (*ABC*, Section 11.6) between two players, A and B. When player A serves, he has probability P1 of winning the point directly; similarly player B has probability P2 of winning directly from his service. If a rally takes place, A has probability P3 of winning it. This section of the program determines who wins each single point; run it several times, and print out the winner each time. Examine the effect of changing P1, P2 and P3.

Using the standard system of tennis scoring, the next section of the program should simulate a complete game, printing out the score after each point and recording the winner; all services in one game are made by the same player. For the next game, the other player serves. Build up in this way the progress of a set, the winner being the player who reaches 6 games first *unless* the other player has by then won 5; in this case the set goes to 7–5 or 7–6 before it is decided.

Run the complete program several times, varying P1, P2 and P3, and printing suitable descriptive output.

2 Modify Program 9.1, *Booking Office Simulation*, to allow for a third category of customer, who pays by credit card; allow 60 seconds for this transaction.

3 Add to Program 9.1 instructions to print out the time, and the length of the queue, as each customer arrives.

4 Modify Program 9.1 so that a second service-point is opened when the queue-length exceeds some prescribed value Q; the two service-points continue operating until one has no customers, and it then closes.

5 Modify Program 7.1, *Card Collecting Simulation*, to include an element of 'rarity', for example collecting a set of stamps of a foreign country when the smaller-priced stamps are easier to find than the larger-priced ones.

6 Rewrite Program 9.1, *Booking Office Simulation*, to allow the service times to have normal distributions with the means 15 s and 45 s previously used; set the standard deviations equal to 3 s and 7 s respectively.

Extend question 2 in a similar way, making the credit card transaction time normally distributed with mean 60 and standard deviation 10 s. [Refer to Section 9.5 on normal variables with general values for mean and variance.]

7   Write a program which:
 i) generates a random sample from a named distribution (using a subroutine);
 ii) generates N1 such random samples and finds their mean;
 iii) repeats this N times;
 iv) forms a frequency table of the resulting N means;
 v) plots a histogram of these means.

8   Run the program described in question 7 with a uniform distribution on the interval $(0, 1)$ as the named distribution. Run simulations of $N = 100$ samples with N2 taking the values (i) 2, (ii) 6, (iii) 12. Do the three frequency tables and histograms produced resemble those of normal distributions? [See Exercises 10.7, question 10, for a goodness-of-fit test for a normal distribution.]

9   Repeat question 8 with an exponential distribution whose mean is 1 as the named distribution.

10   Run Algorithm 9.4, *Random – Poisson*, (i) with $\lambda = 0.75$, (ii) with $\lambda = 4$. Repeat each several times, build up frequency distributions, and compare them with the Poisson distributions with the same values of λ (see Section 8.2.2).

11   Run Algorithm 9.5, *Random – Geometric*, (i) with $P = 0.25$, (ii) with $P = 0.6$. Repeat each several times, build up frequency distributions, and compare them with the geometric distributions with the same values of P (see Section 8.2.3).

12   *Unbiased estimation of variance.* Generate three random samples from a normal distribution with mean 5 and variance 9. Find the mean of these three, and also their variance (i) using one of the methods in Chapter 6 with $(N-1)$, i.e. 2, as divisor; (ii) replacing $(N-1)$ by N, i.e. 3. Repeat this 100 times, and build up a frequency distribution for the mean of the three, and for the variance calculated by both methods. Calculate the mean and variance in each of these three frequency distributions.

13   Box and Muller's method of generating normal variables (Section 9.5.2) can be modified to take account of non-independence when using congruential generators, as follows:

```
530  W1 = RND(1)
540  W2 = INT(10*RND(1))
542  FOR I = 1 TO W2
544  T = RND(1)
546  NEXT I
548  W2 = T
```

Make these changes in Algorithm 9.7, *Random–Normal (Box–Muller)*, run this and the unmodified program several times and compare results.

14   Write a program to draw random samples from a gamma distribution, as described in Section 9.7.2. Run it with (i) $\alpha = 5, \beta = 3$; (ii) $\alpha = 6, \beta = 7$. Also use it to draw random samples from a χ^2 distribution with 6 degrees of freedom.
  In each case take a large number of samples and summarise the results in a grouped frequency table.

15 Write a flow diagram for Program 9.1, *Booking Office Simulation*, including the modification described in question 2.

Make any changes needed to include the random variables specified in question 6.

16 Marsaglia and Bray have proposed a method of generating normal random variables which avoids the use of trigonometric functions (computers are likely to be slow in evaluating these functions). A rejection method is used. The procedure is:

1) generate two $U(0, 1)$ random variables u_1, u_2;
2) calculate $w_1 = 2u_1 - 1$, $w_2 = 2u_2 - 1$;
3) calculate $w = w_1^2 + w_2^2$; if $w \geq 1$ return to (1), otherwise continue;
4) calculate $x_1 = w_1(-2(\ln w_1)/w_1)^{1/2}$, $x_2 = w_2(-2(\ln w_2)/w_2)^{1/2}$. These are two independent standard normal variables.

Write an algorithm for this procedure.

10 Significance tests and confidence intervals

10.1 Introduction

We now consider two methods which the statistician may use in order to make inferences about a population based on observations from a sample that he has collected: significance testing (*ABC*, Chapters 12, 16, 18); and estimation by calculating confidence intervals (*ABC*, Chapter 17). As a basis for a significance test, we set up a null hypothesis and make suitable assumptions about our data; then we obtain a suitable statistic, such as a t-statistic, or a χ^2-statistic, whose sampling distribution is known provided that our assumptions are justified. The value of the statistic is then calculated for our particular set of data, and this is compared with tables of the theoretical sampling distribution to complete the test. After we have calculated obvious summary statistics of a set of data, such as the mean and variance, determining the value of a test statistic (such as t) is often a trivial calculation for a computer, and will occupy only a few lines of program.

But the actual outcome of a significance test is just one item of evidence in interpreting a set of data. We must be sure to output, as well as the value of the test statistic, all the relevant information we need to set the significance test in context. The computer can be made to print out critical values or p-values of the sampling distributions of the statistics. We have not written these procedures into our algorithms for test statistics, but we shall indicate how the algorithms that we gave in Chapter 8 for probability functions may be used to produce this information.

When finding a confidence interval for a population parameter, such as a mean, we need an estimate of the parameter, such as the sample mean, and we also need to know what is the sampling distribution of this estimator. In many commonly useful cases, provided that certain standard assumptions hold, the sampling distribution will be a normal or a t distribution, whose mean is the unknown population parameter. When this is so, the confidence interval is given by the estimate plus or minus t standard errors, where t is a factor that is usually found from tables. By using a computer, it is possible to avoid looking up tables, and calculate the factor t to an adequate level of accuracy.

We consider algorithms according to the type of data available: single sample, two samples (paired or unpaired), or frequency data. Since the basic arithmetic for significance testing and for confidence intervals is similar, it is often convenient to consider both processes together for computing purposes, even though they are using the sample data in different ways.

10.2 Single sample analysis

Suppose we have an array of data $X(\cdot)$, containing n items, which forms a random sample from a population. If in this population x is normally distributed, or approximately so, we may use a t-statistic to test whether the mean $\bar{x}$ differs significantly from a specified value μ (*ABC*, Section 16.9), and we may also construct a confidence interval for the true mean μ (*ABC*, Section 17.5). We give output algorithms for both these procedures.

10.2.1 The *t*-test

When a random sample of n items is available from a population, the sample mean being $\bar{x}$ and estimated variance s^2, we may wish to test the null hypothesis that the true population mean is μ_0. This requires the *t*-statistic

$$t = \frac{(\bar{x} - \mu_0)}{\sqrt{s^2/n}},$$

whose sampling distribution is t with $(n-1)$ degrees of freedom. Algorithm 10.1, *Output – t Test on Mean*, prints out the statistic and other relevant information, given the four quantities on the right-hand side of the above formula.

```
500   REM   ** OUTPUT - T TEST ON MEAN **
510   REM   * INPUT: NAME OF VARIATE X$ *
520   REM   * NUMBER OF DATA N, SAMPLE MEAN M *
530   REM   * ESTIMATED VARIANCE V, MEAN ON NULL HYPOTHESIS MØ *
540   REM   * VARIABLES: TØ *

550   LET TØ = (M - MØ)/SQR(V/N)
560   PRINT X$
570   PRINT
580   PRINT "T-STATISTIC (MEAN OF POPULATION)"
590   PRINT
600   PRINT "MEAN ON NULL HYPOTHESIS      "; MØ
610   PRINT
620   PRINT "SAMPLE MEAN                  "; M
630   PRINT
640   PRINT "STANDARD ERROR OF THE MEAN "; SQR(V/N)
650   PRINT
660   PRINT "T-STATISTIC                  "; TØ
670   PRINT
680   PRINT "DEGREES OF FREEDOM           "; N - 1
```

Algorithm 10.1 *Output – t Test on Mean*

Information on the sampling distribution may be obtained by using the algorithms for the t distribution from Chapter 8. The critical value of t needed in a 5% two-tail significance test can be found by inputting $P\emptyset = .975$ into *Inverse t – Approx* (Algorithm 8.10). In general for a two-tail test the probability value $P\emptyset$ which is input is such that the upper tail probability $(1 - P\emptyset)$ equals half the significance level; for a one-tail test $(1 - P\emptyset)$ must be set equal to the significance level. We wish to obtain the probability, or *p*-value, that our particular calculated *t*-value for a sample is exceeded in absolute value. To do this, we input the absolute value, $T\emptyset$, of our t into *t Distribution Fnc* (Algorithm 8.7), and obtain as output $P\emptyset$. Then $1 - P\emptyset$ is the appropriate probability for a one-tail test, and $2(1 - P\emptyset)$ for a two-tail test.

10.2.2   Confidence interval

A 95 % central confidence interval for the true population mean μ is given by the expression (*ABC*, Section 17.5)

$$\bar{x} - \sqrt{s^2/n} \cdot t_{(n-1,0.05)} \le \mu \le \bar{x} + \sqrt{s^2/n} \cdot t_{(n-1,0.05)}.$$

In this expression t stands for the value which is exceeded by the absolute value of a t variable, with $(n-1)$ degrees of freedom, with probability 0.05. Thus both tails of the t variable are taken into account. If $F(t)$ is the cumulative distribution function of the t variable with $(n-1)$ degrees of freedom, then the value of t we require is the solution of $F(t) = 0.975$. Algorithm 10.2, *Output – Confidence Interval for Mean*, calculates the confidence interval for any confidence level, not just for 95 %; it uses *Inverse t – Approx* (Algorithm 8.10) as a subroutine to calculate the appropriate value of t to put in the expression. One could avoid using this subroutine by arranging to insert the appropriate t-value as part of the input.

```
500   REM   ** OUTPUT - CONFIDENCE INTERVAL FOR MEAN **
510   REM   * INPUT: NAME OF VARIATE X$, SAMPLE MEAN M *
520   REM   * ESTIMATED VARIANCE V, DEGREES OF FREEDOM K1 *
530   REM   * NUMBER OF DATA N, CONFIDENCE PROBABILITY KØ *
540   REM   * REQUIRES: INVERSE T (8.10) AT 1500 *
550   REM   * VARIABLES: I1, I2, PØ, S, TØ, W *

560   REM   FIND T MULTIPLIER
570   LET PØ = Ø.5 + KØ/2
580   GOSUB 1500

590   LET S = SQR(V/N)
600   LET I1 = M - TØ*S
610   LET I2 = M + TØ*S
620   LET W = 100*KØ
630   PRINT X$
640   PRINT
650   PRINT W; " % CONFIDENCE INTERVAL FOR MEAN"
660   PRINT "("; I1; ","; I2; ")"
```

Algorithm 10.2 *Output – Confidence Interval for Mean*

If the variance in the population is known, and need not be estimated from the sample (a situation common in textbooks but rare in practice!) the appropriate value from the standard normal distribution replaces t in the expression for the confidence interval (e.g. the value 1.96 if we require a 95 % central confidence interval).

10.3   Paired samples

For this case, we assume that two arrays $X(\cdot)$, $Y(\cdot)$ are given, such that, for all I, $X(I)$ is paired with $Y(I)$. There are N units. This is the situation when two observations, denoted by X and Y, have been recorded on each unit in the sample of N. Significance tests (*ABC*, Section 16.9.1) and confidence intervals are based on the *differences* between the paired values; these differences are assumed to be normally distributed.

To test the null hypothesis that the true means in the two populations X and Y are equal, we begin by forming the array of differences whose typical element is $D(I) = X(I) - Y(I)$. We then proceed as for the single-sample case, with $D(\cdot)$ replacing $X(\cdot)$ and M0 set equal to zero. Algorithm 6.1 or 6.2 may be used to find the mean and estimated variance of the differences $D(\cdot)$. (To use either of these algorithms the elements of the array $D(\cdot)$ have to be put into $X(\cdot)$; it would be wise to preserve the original elements of $X(\cdot)$ by first placing them in $U(\cdot)$, say, so that after the mean and variance of the differences have been determined the original observations may be recovered from $U(\cdot)$ and put back into $X(\cdot)$.) Algorithm 10.1 may be used to give the *t*-statistic for the test on the mean *difference*, and similarly Algorithm 10.2 will give a confidence interval for the mean *difference*. Slight changes in the printed output would of course be needed in doing this.

10.3.1 Wilcoxon signed-rank test

This is a non-parametric test of the null hypothesis that the differences between the pairs of observations have a median of zero (Walpole, 1974). In the first part of Algorithm 10.3, *Wilcoxon Signed Rank*, the absolute values of the differences are found and these are ranked. Tied observations are given average ranks. If any differences are zero, these are not used when calculating the test statistic; this can lead to awkwardness if we try to change the value of N when entering the ranking algorithm and later reset it in case it is needed elsewhere in the program. We may avoid this difficulty by replacing zero differences in the absolute difference vector $X(\cdot)$ by very large numbers. This ensures that the ranks of all the other differences will be the same as they would have been if the zero differences had been discarded. Line 620 of Algorithm 10.3 does this. The number of zero differences is counted by N0, and the number of non-zero differences is then $N2 = N - N0$.

The test statistic T is the sum of the ranks of the N1 elements of $X(\cdot)$ which correspond to *negative* differences. The sampling distribution of T is symmetric about a mean of $N2*(N2 + 1)/4$; tables usually quote only the lower tail of the distribution of T. If the value of T exceeds its mean, we have an upper tail value, and in order to obtain the corresponding value in the lower tail we must subtract it from $N2*(N2 + 1)/2$; in fact this lower tail value is the sum of the ranks of the positive differences. The form of output that we recommend includes both tail values. For large values of N2 the sampling distribution of T is approximately normal, and in such cases a standard normal deviate z may be used for the test; this statistic is calculated in the last few lines of the algorithm.

A possible form of output follows, in which the expressions in italics show where their numerical values should appear.

```
T$ (Title)
```

DIFFERENCES	NUMBER	SUM OF RANKS
SAMPLE 1 < 2	*N1*	*T*
SAMPLE 2 < 1	*N2−N1*	*N2*(N2+1)/2−T*
SAMPLE 1 = 2	*N0*	

```
STANDARD NORMAL DEVIATE Z
(VALID IF NUMBER OF NON-ZERO DIFFERENCES > 19)
```

```
500   REM   ** WILCOXON SIGNED RANK **
510   REM   * INPUT: PAIRED DATA G(•), H(•) *
520   REM   * NUMBER OF UNITS N *
530   REM   * REQUIRES: RANKING (3.3) AT 1500 *
540   REM   * VARIABLES: I, NØ, N1, N2, N3, T, W, W3, Z *
550   DIM W(100), X(100), R(100)

560   REM   CALCULATE AND RANK ABSOLUTE DIFFERENCES
570   LET NØ = Ø
580       FOR I = 1 TO N
590       LET W = G(I) - H(I)
600       IF W <> Ø THEN 640
610       LET NØ = NØ + 1
620       LET X(I) = 1E38
630       GO TO 650
640       LET X(I) = ABS(W)
650       LET W(I) = W
660       NEXT I
670   REM   RANKS RETURNED IN R(•)
680   GOSUB 1500

690   REM   CALCULATE STATISTICS
700   LET T = Ø
710   LET N1 = Ø
720       FOR I = 1 TO N
730       IF W(I) >= Ø THEN 760
740       LET N1 = N1 + 1
750       LET T = T + R(I)
760       NEXT I
770   LET N2 = N - NØ
780   LET N3 = N2*(N2 + 1)/4
790   LET W3 = N3*(N2 + N2 + 1)/6
800   LET Z = (T - N3)/SQR(W3)

810   REM   * WILCOXON STATISTIC T *
820   REM   * LARGE SAMPLE STATISTIC Z *
```

Algorithm 10.3 *Wilcoxon Signed Rank*

10.4   Two unpaired samples

The sample of n_1 observations from one population is stored in the array X(·), and the sample of n_2 observations from the other population in Y(·).

10.4.1   The *t*-test

The commonly used parametric test is a *t*-test, valid when we are sampling from two normal populations. The *t*-statistic for testing the null hypothesis that the true means of the two populations are equal is

$$t = -\frac{\bar{x}_1 - \bar{x}_2}{\sqrt{s^2\left(\dfrac{1}{n_1} + \dfrac{1}{n_2}\right)}}$$

in which $\bar{x}_1, \bar{x}_2$ are the means of the two samples and s^2 is a pooled estimate of variance: the t-test is not valid (at least for small n_1, n_2) if the variances in the two populations are different. Therefore we print out the two estimated variances s_1^2, s_2^2 for the two samples, to ensure that they are not greatly different and that pooling the estimates is legitimate. The pooled estimate of variance is

$$s^2 = \frac{(n_1 - 1)s_1^2 + (n_2 - 1)s_2^2}{n_1 + n_2 - 2}.$$

The sampling distribution of t as defined above is a t distribution with $(n_1 + n_2 - 2)$ degrees of freedom, and the statistic is calculated in Algorithm 10.4, *t-Statistic – Unpaired Samples*.

```
500   REM   ** T STATISTIC - UNPAIRED SAMPLES **
510   REM   * INPUT: MEANS M1, M2; NUMBER OF DATA N1, N2 *
520   REM   * ESTIMATED VARIANCES V1, V2 *
530   REM   * VARIABLES: K1, TØ, V, W *

540   LET K1 = N1 + N2 - 2                          Degrees of freedom
550   LET W = (N1 - 1)*V1 + (N2 - 1)*V2
560   LET V = W/K1
570   LET W = SQR(V/N1 + V/N2)
580   LET TØ = (M1 - M2)/W

590   REM   * OUTPUT: T STATISTIC TØ *
```

Algorithm 10.4 *t-Statistic – Unpaired Samples*

10.4.2   Confidence interval
A confidence interval for the difference between the true means of the two normal populations, in the conditions described in Section 10.4.1, is constructed in the same way as we proceeded in 10.2.2. The difference $(\bar{x}_1 - \bar{x}_2)$ replaces $\bar{x}$; s^2 is calculated from the formula in Section 10.4.1 and then s^2/n in the equation of Section 10.2.2 is replaced by $s^2\left(\dfrac{1}{n_1} + \dfrac{1}{n_2}\right)$; and the appropriate degrees of freedom are $(n_1 + n_2 - 2)$.

10.4.3   Mann–Whitney U test
This is a nonparametric test of the identity of two populations from which independent random samples have been drawn (Walpole, 1974). It can therefore be applied when the normality condition in Section 10.4.1 breaks down. In Algorithm 10.5, *Mann–Whitney Test*, all the data are merged, and the rank of each observation is found using a subroutine. We have suggested using *Ranking* (Algorithm 3.3), but any other ranking subroutine might be used provided that variables are set appropriately before the subroutine is entered. The sum T of the ranks of the observations in the first sample is found; the statistic U upon which the significance test is based is equal to $T - n_1(n_1 + 1)/2$. In fact, U is the number of times members of sample 2 precede members of sample 1 in the joint ranking. The sampling distribution of U is symmetric, with mean $n_1 n_2/2$; but only the lower half of the distribution is given in tables. Thus for a

```
500   REM   ** MANN-WHITNEY TEST **
510   REM   * INPUT: DATA X(•), Y(•) *
520   REM   * SAMPLE SIZES N1, N2; DIM X(•) >= N1 + N2 *
530   REM   * REQUIRES: RANKING (3.3) AT 1500 *
540   REM   * VARIABLES: I, N, N3, T1, UØ, W, Z *
550   DIM R(100)

560      FOR I = 1 TO N2
570      LET X(N1 + I) = Y(I)                      Merge arrays
580      NEXT I
590   LET N = N1 + N2

600   REM   RANK COMBINED DATA
610   GOSUB 1500

620   LET T1 = Ø
630      FOR I = 1 TO N1
640      LET T1 = T1 + R(I)                        Sum ranks of sample 1
650      NEXT I
660   LET UØ = T1 - N1*(N1 + 1)/2                  U-statistic
670   LET N3 = N1*N2
680   LET W = SQR(N3*(N + 1)/12)
690   LET Z = (UØ - N3/2)/W                        Large sample statistic

700   REM   * OUTPUT: MANN-WHITNEY STATISTIC UØ *
710   REM   * SAMPLE 1 RANK SUM T1 *
720   REM   * LARGE SAMPLE STATISTIC Z *
```

Algorithm 10.5 *Mann–Whitney Test*

particular set of data we may need the statistic $(n_1 n_2 - U)$ for comparison with the values in tables, and so we suggest outputting the U-statistic for each tail: these two values may be regarded as corresponding to the two samples. For large samples of data—in fact, when n_1 or n_2 is greater than 19—a normal approximation to the distribution of the U-statistic can be used, and the standard normal deviate Z for this is calculated in the final section of Algorithm 10.5.

When observations are tied in rank, an average rank is given to each observation affected: *Ranking* (Algorithm 3.3) does this automatically. There is a correction to the variance used in the normal approximation which takes account of ties; its effect, however, is negligible unless the number of ties is large and so we have not incorporated it. An example of the type of output we would recommend is as follows; expressions in italics show where their numerical values should appear.

T$ (Title)

	SAMPLE 1	SAMPLE 2
SUM OF RANKS	*T1*	*N*(N+1)/2−T1*
SAMPLE SIZE	*N1*	*N2*
U-STATISTIC	*UØ*	*N1*N2−UØ*

STANDARD NORMAL DEVIATE *Z*
(VALID IF N1 OR N2 > 19)

10.5 Frequency data

There are two main types of test required with frequency data, both of them based on the chi-squared distribution.

10.5.1 Chi-squared goodness-of-fit tests

We sometimes wish to test whether a set of data could reasonably have arisen as a sample from a particular distribution. One way to do this is to group the observed data into classes, find the *observed* frequency in each class, and find also the frequency *expected* in each class if the data really do follow the particular distribution we propose for them. We compare the corresponding observed and expected frequencies using a statistic that follows a χ^2 distribution (*ABC*, Section 18.5).

```
500  REM  ** CHI-SQUARED GOODNESS-OF-FIT **
510  REM  * INPUT: OBSERVED FREQUENCIES O(·) *
520  REM  * EXPECTED FREQUENCIES E(·), NUMBER OF CLASSES C *
530  REM  * NUMBER OF PARAMETERS ESTIMATED FROM DATA N9 *
540  REM  * VARIABLES: I, K1, T1, T2, W, W1, W2, X2 *
550  DIM J(15)

560  LET T1 = 0
570  LET T2 = 0
580      FOR I = 1 TO C                      Sum observed and
590      LET T1 = T1 + O(I)                  expected frequencies
600      LET T2 = T2 + E(I)
610      NEXT I
620  IF ABS(T1 - T2) < 0.5 THEN 650          Test near-equality
630  PRINT "SUM OF OBSERVED FREQ = "; T1
640  PRINT "SUM OF EXPECTED FREQ = "; T2

650  REM  CALCULATE CHI-SQUARED STATISTIC
660  LET X2 = 0
670      FOR I = 1 TO C
680      LET W1 = E(I)
690      LET W2 = O(I) - W1
700      IF W1 >= 5 THEN 720                  Test for small expectation
710      PRINT "SMALL EXPECTED FREQUENCY E("; I; ") = "; E(I)
720      LET W = W2*W2/W1
730      LET J(I) = W                         Store component of χ²
740      LET X2 = X2 + W                      Accumulate χ²
750      NEXT I
760  LET K1 = C - 1 - N9                      Degrees of freedom

770  REM  * OUTPUT: CHI-SQUARED VALUE X2 *
780  REM  * DEGREES OF FREEDOM K1 *
790  REM  * CHI-SQUARED COMPONENTS J(·) *
```

Algorithm 10.6 *Chi-squared Goodness-of-fit*

In Algorithm 10.6, *Chi-squared Goodness-of-fit*, the observed frequencies in the classes are input in an array $O(\cdot)$. The expected frequencies $E(\cdot)$ are calculated using the appropriate probability function. The totals of observed and expected frequencies should be the same; this is tested in line 620, but since the sums may not be exactly equal due to rounding errors in calculating the expected frequencies, we test whether the difference T1 − T2 is small rather than exactly zero. An error message is printed if there is a discrepancy between the totals. Some—usually all—of the parameters of the theoretical distribution are made to equal their estimates calculated from the sample data. Thus when the theoretical distribution is the Poisson, its parameter λ, which equals the mean, will often be made equal to the mean of the sample data; similarly if the theoretical distribution is the normal, its parameters μ and σ^2, which equal the mean and variance respectively, will often be made equal to the mean and estimated variance from the sample. If we set up these correspondences for parameters, we affect the degrees of freedom of the sampling distribution of the χ^2-statistic, and hence affect the interpretation of the results. The general rule is that the number of degrees of freedom is equal to the 'number of classes, minus one for the total, minus one for each parameter estimated' (*ABC*, Section 18.6.1). The number of parameters estimated from the data is thus necessary information; it is input as N9.

The χ^2-statistic equals $\Sigma(O_i - E_i)^2/E_i$, and is calculated in lines 650–750 of Algorithm 10.6. The component of χ^2 deriving from each cell is stored in an array J(·); it can be helpful to inspect these values when interpreting a large total χ^2. In fact this statistic is only *approximately* distributed as χ^2, and the approximation is poor if too many of the expected cell frequencies are small. A rule of thumb is that no expected frequency should be less than 5; however, this is really being rather too strict, and the topic is discussed by Snedecor and Cochran (1980). We make the algorithm print a warning if any expected frequency falls below a chosen value, which we have set at 5 (although it is very easy to alter this choice). If expected frequencies in some classes are too small, adjacent classes should be merged to produce classes with sufficiently large expected frequencies.

A reasonable form of output is as follows, where we have inserted the algorithm variable names (in italics) in the appropriate places.

```
OBSERVED      EXPECTED      X2 COMPONENT

O(1)          E(1)          J(1)
O(2)          E(2)          J(2)
  .             .             .
  .             .             .
  .             .             .
O(C)          E(C)          J(C)


X2 STATISTIC           X2
DEGREES OF FREEDOM     K1
```

The critical value needed for an $\alpha\%$ significance test may be obtained by putting P0 = 1 − α in *Inverse Chi-Squared – Approx* (Algorithm 8.11), using the appropriate degrees of freedom. The probability of the χ^2-statistic, found for a particular set of data, being exceeded is obtained by input of X2 into *Chi-Squared Distribution Fnc* (Algorithm 8.8); the required probability is 1 − P0, where P0 is the output from this algorithm.

10.5.2   Chi-squared contingency table tests

When data have been classified in the form of a two-way contingency table, the usual null hypothesis for testing is that the two classifications are independent of one another. On this assumption the expected frequencies in the cells of the table may be derived

```
500   REM   ** CHI-SQUARED CONTINGENCY TABLE **
510   REM   * INPUT: OBSERVED FREQUENCIES O(•,•) *
520   REM   * NUMBER OF ROWS N1, OF COLUMNS N2 *
530   REM   * VARIABLES: I, J, K1, T, T1, W, W1, W2, X2 *
540   DIM E(5,5), C(5), J(5,5), R(5)

550   REM   CALCULATE TABLE TOTALS
560       FOR J = 1 TO N2
570       LET C(J) = 0
580       NEXT J
590   LET T = 0
600       FOR I = 1 TO N1
610       LET T1 = 0
620           FOR J = 1 TO N2
630           LET W = O(I,J)
640           LET T1 = T1 + W
650           LET T = T + W                         Grand total
660           LET C(J) = C(J) + W                   Column total
670           NEXT J
680       LET R(I) = T1                             Row total
690       NEXT I

700   REM   CALCULATE EXPECTED VALUES E(I,J)
710       FOR I = 1 TO N1
720       LET W = R(I)/T
730           FOR J = 1 TO N2
740           LET W1 = W*C(J)
750           LET E(I,J) = W1
760           IF W1 >= 5 THEN 780
770           PRINT "SMALL EXPECTED FREQUENCY E("; I; ","; J; ") = "; E(I,J)
780           NEXT J
790       NEXT I

800   REM   CALCULATE CHI-SQUARED STATISTIC
810   LET X2 = 0
820       FOR I = 1 TO N1
830           FOR J = 1 TO N2
840           LET W1 = E(I,J)
850           LET W2 = O(I,J) - W1
860           LET W = W2*W2/W1
870           LET J(I,J) = W
880           LET X2 = X2 + W
890           NEXT J
900       NEXT I
910   LET K1 = (N1 - 1)*(N2 - 1)

920   REM   * OUTPUT: CHI-SQUARED VALUE X2 *
930   REM   * DEGREES OF FREEDOM K1 *
940   REM   * COMPONENTS OF CHI-SQUARED J(•,•) *
```

Algorithm 10.7 *Chi-squared Contingency Table*

from the marginal totals of the observed table. The form of the test statistic is the same as for a goodness-of-fit test, namely $\Sigma (O_i - E_i)^2/E_i$ (*ABC*, Section 18.7).

Algorithm 10.7, *Chi-squared Contingency Table*, first calculates totals and expected values, and then proceeds as in the goodness-of-fit test of Section 10.5.1, a warning message being printed if any expected frequency is small. Output could usefully contain tables of χ^2 components, the observed and expected frequencies, the value of the χ^2-statistic and the degrees of freedom. Critical values and *p*-values may be obtained as explained in Section 10.5.1.

10.6   Test data for Algorithms

Algorithm 10.3, *Wilcoxon Signed Rank*. (i) Input: $G(\cdot) = (2, 9, 7, 4, 4, 8)$; $H(\cdot) = (4, 8, 8, 7, 8, 6)$, $N = 6$. Output: $T = 16$, $Z = 1.15311$. (ii) Input: $G(\cdot) = (2, 9, 7, 4, 4, 8, 5, 5)$; $H(\cdot) = (4, 8, 8, 7, 8, 6, 5, 6)$; $N = 8$. Output: $T = 21.5$, $Z = 1.26773$.
Algorithm 10.4, *t-Statistic, Unpaired Samples*. Input: $M1 = 5$, $M2 = 3$, $N1 = 10$, $N2 = 15$, $V1 = 4$, $V2 = 7$. Output: $T(23) = 2.02963$.
Algorithm 10.5, *Mann–Whitney Test* (i) Input: $X(\cdot) = (2, 3, 7, 10)$, $Y(\cdot) = (5, 6, 9, 12, 13)$, $N1 = 4$, $N2 = 5$, Dimension of $X(\cdot) \geq 9$. Output: $U\emptyset = 5$, $T1 = 15$, $Z = -1.22474$. (ii) Input: $X(\cdot) = (2, 3, 7, 10, 12)$, $Y(\cdot) = (5, 6, 9, 12, 13, 15, 19)$, $N1 = 5$, $N2 = 7$, Dimension of $X(\cdot) \geq 12$. Output: $U\emptyset = 8.5$, $T1 = 23.5$, $Z = -1.46160$.
Algorithm 10.6, *Chi-squared Goodness-of-fit*. Input: $O(\cdot) = (4, 10, 8, 13, 16)$, $E(\cdot) = (5, 11, 12, 15, 8)$, $C = 5$, $N9 = 0$. Output: $X2 = 9.89091$, $K1 = 4$.
Algorithm 10.7, *Chi-squared Contingency Table*. Input: $O(\cdot, \cdot) = (10, 50, 15, 25)$, $N1 = 2$, $N2 = 2$. Output: $X2 = 5.55556$, $K1 = 1$, $J(\cdot, \cdot) = (1.6\dot{6}, 0.5\dot{5}, 2.5, 0.8\dot{3})$.

10.7   Exercises

1   Write a program which:
1) generates a random sample of size N2 from a normal distribution with mean M∅ and variance V∅;
2) calculates the sample mean and variance;
3) calculates a 95 % confidence interval for the true mean, based on the sample data (i.e. *not* using the known values M∅, V∅);
4) repeats N times, and counts the number of times N3 that M∅ is actually contained in the confidence interval;
5) prints out N3/N.

Run the program with $M\emptyset = 10$, $V\emptyset = 4$, $N2 = 10$, $N = 100$.
Modify step (3) to calculate a 99 % confidence interval, and run the program with the same values of M∅, V∅ and N2; take $N = 200$.

2   Consider a sample of pairs of values (x_i, y_i), $i = 1, 2, \ldots n$, from a bivariate normal distribution. On the null hypothesis that the true population correlation coefficient is 0, the statistic $T = r\sqrt{(n-2)}/\sqrt{1 - r^2}$ follows a *t* distribution with $(n - 2)$ degrees of freedom. Write a program which, when given *n* pairs of values (x, y),
1) prints a scattergram of the data (see Algorithm 5.5);

2) calculates the sums-of-squares-and-products matrix of the data (see Algorithm 6.3 or 6.4);

3) prints out the values of (i) the correlation coefficient, (ii) the *t*-statistic described above, (iii) the number of degrees of freedom.

3 The *sign test* (*ABC*, Section 12.2.2) requires N pairs of values (x, y). It attaches a + sign to each pair in which $x > y$, a $-$ sign when $x < y$, and ignores pairs in which $x = y$: suppose there are N1 of the latter type. On the null hypothesis that there is no difference between the medians of x and y in the population, the number of plus signs should be binomially distributed with the parameter $p = \frac{1}{2}$ and $n = N - N1$.

Write a program which, when given N pairs of values (x, y),
1) tests whether each pair should be allocated a + or a $-$ sign, or be ignored;
2) counts the number of plus signs, R, and the number of pairs ignored, N1;
3) finds the probability P0 of R in the appropriate binomial distribution (see Algorithm 8.1);
4) prints out N, N1, R and P0.

4 Write a program to carry out a goodness-of-fit test on a set of observations which are said to be from a binomial distribution with a given *n* and *p*. Include instructions for combining frequencies at the beginning and/or end of the distribution so as to avoid expected frequencies being too small (see Section 10.5.1).

5 Generate a set of 100 observations from a binomial distribution with $n = 12$, $p = 0.25$ (using Algorithm 9.3) and use the program written for question 4 to make a goodness-of-fit test on them.

Print out a frequency table showing 'observed' (i.e. generated) frequencies, expected frequencies, χ^2 value and number of degrees of freedom.

6 Modify Algorithm 10.7, *Chi-squared Contingency Table*, to include a check that all frequencies input are integers and that none are negative.

7 Use Algorithm 10.6, *Chi-squared Goodness-of-fit*, to examine
a) the hypothesis that a die is 'fair' (properly balanced) if 120 throws gave 15 ones, 25 twos, 18 threes, 15 fours, 23 fives, 24 sixes;
b) the hypothesis that a set of three coins is 'fair' if 80 tosses of the three gave no heads 10 times, one head 25 times, two heads 34 times and three heads 11 times.

8 *Fitting an exponential distribution to a set of data.* The life-times of 500 items of a particular electrical component have been summarised in the following frequency table.

Time (hours)	Frequency
0– 99	208
100–199	112
200–299	75
300–399	40
400–499	30
500–599	18
600–699	11
700–799	6

1) Write a section of program to estimate the mean of these observations. (The first part of question 8 of Exercises 6.10 will be appropriate.)
2) Continue the program by finding the expected frequencies in an exponential distribution with the same mean (i.e. whose parameter is 1/mean).
3) Carry out a goodness-of-fit test of these expected frequencies to those observed.
4) Print out the observed values in each interval, the corresponding expected values, the χ^2-value from the test and its number of degrees of freedom.

9   A geometric distribution can be computationally awkward to fit to data because its expected frequencies fall off very slowly. Construct a program for a goodness-of-fit test such that the class-intervals used in the test have approximately equal expected frequencies.

Use the program to test the hypothesis that the following observations X came from a geometric distribution with $p = \frac{1}{10}$ (see Section 8.2.3).

Value of X	1	2	3	4	5	6	7	8	9	10	11	12	13	14	15
Frequency	14	10	10	6	5	9	7	2	2	3	5	3	2	2	1

Value of X	16	17	18	19	20	21	22	23	24	25	29	39	Total
Frequency	4	3	2	1	1	3	2	1	3	2	1	1	105

10†   *Fitting a normal distribution to a set of data.* Write a program similar to that in question 8 to test the goodness-of-fit of a set of data to a normal distribution. (Note that now the variance of the data will need calculating, as well as the mean.) Use it to test whether a normal distribution can be fitted satisfactorily to the following observations on the weights of crop (grams) harvested from 200 samples of plants in a field of wheat.

Crop (g)	60–79	80–99	100–119	120–139	140–159	160–179	180–199
Frequency	6	14	67	87	19	5	2

11   (i) Write a program to test the goodness-of-fit of a set of data to a Poisson distribution, the data being given in a frequency table. Make provision for *either* (a) the mean of the distribution to be specified as input; *or* (b) the mean of the data to be calculated and used as the mean in the fitted distribution.
(ii) Write a program to test the fit of data to a Poisson distribution by the 'Index of Dispersion': if N is the total number of observations in the set of data $\{x_i\}$, then $\sum_{i=1}^{N} (x_i - \bar{x})^2 / \bar{x}$ is (approximately, for large N) a χ^2 variable with $(N-1)$ degrees of freedom.

12   A town records the number of motor accidents on its roads each week; at the end of each week it alters the number on boards displayed on each main road entering the town; these say 'there have been (X) accidents in this town this year'. The numbers X exhibited in successive weeks during a year were 3, 5, 6, 8, 8, 11, 12, 13, 15, 17, 20, 21, 25, 25, 26, 31, 34, 36, 38, 43, 43, 47, 49, 50, 58, 61, 62, 62, 64, 66, 66, 69, 73, 75, 80, 82, 82, 85, 86, 87, 87, 89, 91, 94, 94, 97, 102, 103, 107, 107, 109, 110.

Write a program to extract the *weekly* figures, and to test whether these follow a Poisson distribution (by either of the methods suggested in question 11).

11 Regression

11.1 Introduction

In many investigations we can propose that there exists a linear relationship between variables, and then we use this relation to predict the value of one variable given the value of the others. For a simple two-variable situation we can do this by fitting a regression line to the measurements taken on the two variables, or some transformation of them (*ABC*, Chapter 21). When more than two variables are involved, the linear statistical model is rather more complicated (Draper and Smith, 1981). We discuss algorithms for both these situations in this chapter.

11.2 Fitting a line to pairs of values

The most common situation is when we have records of pairs of values (x, y), representing pairs of measurements on the same experimental units; we wish to fit a straight line to these when plotted on a graph, and to use the line in predicting one variable from the other. We shall assume that x is the independent variable and y the dependent variable, and shall obtain what is called the 'regression of y on x'; we will then use it to predict y from x. The estimated line takes the form

$$y = a + bx,$$

where $b = \dfrac{\Sigma(x - \bar{x})(y - \bar{y})}{\Sigma(x - \bar{x})^2}$ and $a = \bar{y} - b\bar{x}$, the sums being calculated over all the n pairs of data values available. The major part of the calculation, to derive the sums of squares and products, is carried out using one of the algorithms of Section 6.4; Algorithm 11.1, *Linefit*, assumes that these quantities have been calculated.

Corresponding to each *observed* value y_i ($i = 1, 2, \ldots n$) of the predicted or dependent variable there is a *fitted* value $a + bx_i$. The difference between these two values is called a **residual**:

$$r_i = y_i - a - bx_i.$$

The sizes of these residuals reflect how well the line fits the data: the smaller the residuals the better the fit. Residuals can give very useful information, and should be calculated as a matter of course when fitting a line by computer (although, unfortunately, the calculations can be prohibitive when fitting by hand). *Linefit* (Algorithm 11.1) calculates the estimates of the regression coefficients, and also the residuals, which are stored in $R(\cdot)$. The sum of squares of residuals Σr_i^2 is calculated and stored in T, and $T/(n - 2)$ estimates the variance of the deviations of y about the fitted line.

A suggested form of output is as follows; expressions are shown in italics in the places where their numerical values should appear.

```
FITTED LINE IS Y = A + B*X

COEFFICIENT    ESTIMATE    ST. ERROR

      A           A        SQR(W1*(1/N+M1*M1/S(1,1)))
      B           B        SQR(W1/S(1,1))
```

SOURCE	SS	DF	MS
REGRESSION	$S(2,2)-T$	1	$S(2,2)-T$
RESIDUAL	T	$N-2$	$W1$
TOTAL	$S(2,2)$	$N-1$	

OBSERVED	FITTED	RESIDUAL
$Y(1)$	$F(1)$	$R(1)$
$Y(2)$	$F(2)$	$R(2)$
$\vdots$	$\vdots$	$\vdots$
$Y(N)$	$F(N)$	$R(N)$

Algorithm 11.2, *Output – Linefit*, aims to produce this type of output. The standard error of A is $\sigma \sqrt{\left(\dfrac{1}{n}+\dfrac{\bar{x}^2}{\Sigma(x_i-\bar{x})^2}\right)}$ and of B is $\sigma \sqrt{\left(\dfrac{1}{\Sigma(x_i-\bar{x})^2}\right)}$, where σ^2 is estimated by $\Sigma r_i^2/(n-2)$. The residual sum of squares in the analysis of variance table equals Σr_i^2 and the total sum of squares is $\Sigma(y_i-\bar{y})^2$; the regression sum of squares equals the difference between these two.

```
500   REM   ** LINEFIT **
510   REM   * INPUT: NUMBER OF UNITS N *
520   REM   * INDEP VARIABLE X(•), DEP VARIABLE Y(•) *
530   REM   * MEANS M1, M2; SS AND SP MATRIX S(•,•) *
540   REM   * VARIABLES: A, B, I, T *
550   DIM F(100), R(100)

560   REM   CALCULATE SLOPE AND INTERCEPT
570   LET B = S(1,2)/S(1,1)
580   LET A = M2 - B*M1

590   REM   CALCULATE RESIDUALS AND RESIDUAL SS
600   LET T = 0
610      FOR I = 1 TO N
620      LET F(I) = A + B*X(I)
630      LET R(I) = Y(I) - F(I)
640      LET T = T + R(I)*R(I)
650      NEXT I

660   REM   * OUTPUT: INTERCEPT A, SLOPE B *
670   REM   * FITTED VALUES F(•), RESIDUALS R(•) *
680   REM   * SUMS OF SQUARES: TOTAL S(2,2), RESIDUAL T *
```

Algorithm 11.1 *Linefit*

```
500   REM  ** OUTPUT - LINEFIT **
510   REM  * INPUT: OUTPUT FROM 11.1 *
520   REM  * ROUNDING PARAMETER FØ *
530   REM  * TITLE OF DATA T$ *
540   REM  * VARIABLES: W, W1, W2, W3, W4 *

550   DEF FNR(X) = INT(X*(10↑FØ) + 0.5)/(10↑FØ)
560   PRINT
570   PRINT T$
580   PRINT "FITTED LINE IS Y = A + B*X"
590   PRINT
600   LET W1 = T/(N - 2)
610   LET W2 = SQR(W1/S(1,1))
620   LET W3 = 1/N + M1*M1/S(1,1)
630   LET W4 = SQR(W1*W3)
640   PRINT "COEFFICIENT....ESTIMATE...ST. ERROR"
650   PRINT ".........A"; TAB(15); FNR(A); TAB(26); FNR(W4)
660   PRINT ".........B"; TAB(15); FNR(B); TAB(26); FNR(W2)
670   PRINT
680   PRINT "SOURCE.......SS.........DF...MS"
690   PRINT
700   LET W = FNR(S(2,2) - T)
710   PRINT "REGRESSION"; TAB(13); W; TAB(23); "1"; TAB(28); W
720   PRINT "RESIDUAL"; TAB(13); FNR(T); TAB(23); N - 2; TAB(28); FNR(W1)
730   PRINT "TOTAL"; TAB(13); FNR(S(2,2)); TAB(23); N - 1
740   PRINT
750   PRINT
760   PRINT "OBSERVED.....FITTED.....RESIDUAL"
770       FOR I = 1 TO N
780       PRINT FNR(Y(I)); TAB(13); FNR(F(I)); TAB(23); FNR(R(I))
790       NEXT I
```

Algorithm 11.2 *Output – Linefit*

11.3   Residuals

The statistical model underlying the fit of a straight line to pairs of data values makes an assumption about the 'error' component of each observation, i.e. the discrepancy between the observation and its expected value: it assumes that this component is a random sample from an error distribution whose mean is zero and variance fixed (constant) but unknown.

In fact we cannot know these error components, but the residuals, which are the discrepancies between observed and fitted values, may be taken as a good approximation to them, particularly if we have a large amount of data. One constraint on the residuals that we should note is that their sum is zero, because of the way they are calculated. But otherwise we hope they have the properties of a random sample. If they do not, we may take it as evidence that the assumptions underlying the model are not being satisfied.

The most informative plot is usually thought to be that of residuals against fitted values. We may use *Output – Scattergram* (Algorithm 5.5) with $R(\cdot)$ put into $Y(\cdot)$ and a vector of fitted values put into $X(\cdot)$, to provide the plot. We hope to see a random scatter over the plane; any evidence of pattern suggests a deviation from assumptions. In particular, a wedge-shaped collection of points suggests that the variance is not staying constant but is changing as the level of response y changes. If the points lie on a curve with the residuals containing long runs of positive or negative signs, then this suggests that a curve should be fitted rather than a straight line. Odd points that are remote from the main cluster of points suggest that values may have been misrecorded.

If the data were collected at successive times, or at regular points along a strip of ground in a field experiment, the residuals should be plotted against time or distance to see whether there is evidence of association. With time data it is also worth plotting the residual of the point recorded at time t against the residual of the point recorded at time $t + 1$, for the complete set of points, to check if successive errors are independent or not. Again, *Output – Scattergram* (Algorithm 5.5) might be used for the plot.

Readers will find in Draper and Smith (1981) more information on the interpretation of residuals and on what to do if the assumptions fail.

11.4*   Multiple regression

Sometimes we wish to predict a random variable Y not just from one variable X but from a set of variables $X_1, X_2, \ldots X_c$. For example, we may wish to predict the mass of wood in a tree from measurements of its height, girth and age. The X variables are often called independent variables, and the Y variable the dependent variable. The procedure is called the regression of Y on $X_1, X_2, \ldots X_c$.

We give an algorithm *Multiple Regression* (11.3) which estimates the intercept a and the regression coefficients $b_1, b_2, \ldots b_c$ in the regression equation

$$y = a + b_1 x_1 + b_2 x_2 + \ldots + b_c x_c,$$

using least-squares estimation. The algorithm also calculates the standard errors of the estimates, the sums of squares for an analysis of variance, and the fitted values and residuals. Multiple regression is a powerful tool but it is often misused. We recommend that the reader who has not met the technique before should read an introductory description such as is given in Snedecor and Cochran (1980), and should follow that by reading more specialised books such as Draper and Smith (1981) and Chatterjee and Price (1977).

We assume that the C independent variables $(X_1, X_2, \ldots X_C)$ and the dependent variable Y have been measured on each of N units. The values of the independent variables are put in an $N \times C$ matrix $X(\cdot, \cdot)$ and the values of the dependent variable are put in a vector $Y(\cdot)$ with N elements. When the data are input it is probably most convenient to enter for each unit the values:

> (unit number) y x_1 x_2 $\ldots$ x_C.

The algorithm begins by making $Y(\cdot)$ the $(C + 1)$th column of the matrix $X(\cdot, \cdot)$, and then enters the subroutine *Multivar SS and SP (Deviation Method)* (Algorithm A.5), which is in Appendix B. This subroutine calculates the vector of means $M(\cdot)$ and the matrix of sums of squares and products $S(\cdot, \cdot)$. (Note that since the matrix is symmetric, only the lower triangle need be stored.) The subroutine is a generalisation to many variables of Algorithm 6.3, *SS and SP (Deviation Method)*.

```
500   REM   ** MULTIPLE REGRESSION **
510   REM   * INPUT: NUMBER OF UNITS N *
520   REM   * DEPENDENT VARIABLE Y(•) *
530   REM   * N UNITS * C INDEP VARIABLES, X(•,•) *
540   REM   * REQUIRES: LINEAR SOLVER (A.6) AT 1500 *
550   REM   * MULTIVAR SS AND SP (A.5) AT 2500 *
560   REM   * VARIABLES: A, C1, I, I1, J, S2, S3, T, W, W1 *
570   DIM F(100), R(100), S(60)

580   REM   CALCULATE SS AND SP
590   LET C1 = C + 1
600      FOR I = 1 TO N
610      LET X(I,C1) = Y(I)
620      NEXT I
630   GOSUB 2500
640   LET S3 = Q(C1,C1)                                          Total SS

650   REM   SOLVE NORMAL EQUATIONS
660      FOR I = 1 TO C1
670      LET I1 = I*(I - 1)/2
680         FOR J = 1 TO I
690         LET S(I1 + J) = Q(I,J)
700         NEXT J
710      NEXT I
720   GOSUB 1500
730   LET W = 0
740      FOR I = 1 TO C
750      LET W = W + B(I)*M(I)
760      NEXT I
770   LET A = M(C1) - W                                          Intercept

780   REM   CALCULATE RESIDUALS AND RESIDUAL SS
790   LET S2 = 0
800      FOR I = 1 TO N
810      LET T = A
820         FOR J = 1 TO C
830         LET T = T + B(J)*X(I,J)
840         NEXT J
850      LET F(I) = T
860      LET W = Y(I) - T
870      LET R(I) = W
880      LET S2 = S2 + W*W
890      NEXT I
900   LET W1 = S2/(N - C - 1)

910   REM   * OUTPUT: REGRESSION COEFFICIENTS B(•), INTERCEPT A *
920   REM   * SUMS OF SQUARES: RESIDUAL S2, TOTAL S3 *
930   REM   * FITTED VALUES F(•), RESIDUALS R(•) *
940   REM   * INVERSE MATRIX IN S(•), C INDEP VARIABLES *
```

Algorithm 11.3 *Multiple Regression*

The S matrix may be considered as partitioned into submatrices:

$$\begin{bmatrix} S_1 & k \\ k^T & d \end{bmatrix}.$$

If we write m_i ($i = 1, 2, \ldots C$) for the mean of x_i and m_{C+1} for the mean of y, then S_1 is a $C \times C$ matrix whose i, jth element is $\sum_{p=1}^{n} (x_{ip} - m_i)(x_{jp} - m_j)$, k is a vector whose ith element is $\sum_{p=1}^{n} (x_{ip} - m_i)(y_p - m_{C+1})$ and d is $\sum_{p=1}^{n} (y_p - m_{C+1})^2$. The normal equations, whose solution is the vector of regression coefficients b, are given by

$$S_1 b = k.$$

Algorithm 11.3, *Multiple Regression*, uses another subroutine *Linear Solver* (Algorithm A.6) to solve the equations using a Cholesky decomposition; the solutions are put in the array B($\cdot$). The subroutine also calculates the inverse matrix S_1^{-1} which it puts in place of S: this is needed for calculating the standard errors of the estimates. In order to economise on storage, S is in fact stored as a one-dimensional array.

The algorithm proceeds to calculate the intercept A, the fitted values F($\cdot$) and the residuals R($\cdot$). Earlier in the algorithm, the sum of squares of deviations of y, which is called the **total sum of squares** in the analysis of variance, is stored in S3. The **residual sum of squares** S2 is calculated by summing the squares of the residuals.

Algorithm 11.2, *Output – Linefit*, may easily be adapted to provide an appropriate output. The precise variables used in the list of estimates and the analysis of variance are shown in the output layout given below. The standard error of A is rarely of interest and is not included in the layout. If it should be required it may be calculated from the expression $\sigma \sqrt{m^T S m}$, where m is the vector of variate means, S is the inverse matrix described below and σ is replaced by SQR(W1).

```
COEFFICIENT      ESTIMATE      ST. ERROR

A                A
B(1)             B(1)          SQR(W1*S(1))
 .                .                .
 .                .                .
 .                .                .
B(C)             B(C)          SQR(W1*S(C2))


SOURCE        SS       DF     MS

REGRESSION    S3-S2    C      (S3-S2)/C
RESIDUAL      S2       N-C-1  W1
TOTAL         S3       N-1
```

Note that W1 = S2/(N − C − 1) and that the S matrix, whose diagonal elements are used to calculate the standard errors, is the inverse of the matrix calculated by Algorithm A.5, *Multivar SS and SP*. C2 stands for C*(C + 1)/2.

11.5   Test data for Algorithms

Algorithm 11.1, *Linefit*. Input: N = 5, X($\cdot$) = (3, 6, 7, 11, 13), Y($\cdot$) = (4, 7, 5, 8, 11), M1 = 8, M2 = 7, S(1, 1) = 64, S(1, 2) = 40, S(2, 2) = 30. Output: A = 2, B = 0.625,

$F(\cdot) = (3.875, 5.750, 6.375, 8.875, 10.125)$, $R(\cdot) = (0.125, 1.250, -1.375, -0.875, 0.875)$, $S(2, 2) = 30$, $T = 5$.

Algorithm 11.3, *Multiple Regression*. Input: $N = 5$, $C = 2$, $Y(\cdot) = (4, 7, 5, 8, 11)$. $X(\cdot, \cdot)$ has rows $(3, 10)$, $(6, 8)$, $(7, 5)$, $(11, 4)$, $(13, 3)$. Output: $A = -8.73$, $B(\cdot) = (1.26, 0.93)$. $S2 = 1.73$, $S3 = 30$. $F(\cdot) = (4.40, 6.33, 4.80, 8.93, 10.53)$. $R(\cdot) = (-0.40, 0.66, 0.20, -0.93, 0.46)$.

11.6 Exercises

1 The following data give the reaction times of 10 men, of various ages, to a visual stimulus in a psychological experiment; x is age (in years) and y is reaction time (in milliseconds):

x	37	35	41	43	42	50	49	54	60	65
y	190	197	205	210	218	226	228	230	234	240

Apply Algorithms 11.1 and 11.2, together with one of the algorithms of Section 6.4, to calculate the regression line of y on x and print out the results, including residuals.

2 Write program which
  i) inputs N data in pairs (X, Y);
  ii) plots a scattergram (Algorithm 5.5);
  iii) calculates the SS and SP matrix (Algorithm 6.3);
  iv) prints the correlation coefficient (see Section 6.6);
  v) calculates the regression line of y on x (Algorithm 11.1);
  vi) prints the results of (v) (Algorithm 11.2).

If the program is to be used interactively, insert a STOP instruction at each stage so that the results may be inspected as the program proceeds.

Check the program using the test data for Algorithm 11.1 (Section 11.5).

3 Write a program to:
  i) generate N pairs of data (x, y) according to the model

$$y_i = a + bx_i + e_i \qquad (i = 1, 2, \ldots n)$$

where $x_i = i$ and e_i is a random sample from a normal distribution with mean 0 and variance σ^2;
  ii) estimate the values of the intercept A and slope B in the regression line of Y on X;
  iii) repeat this N1 times and study the distributions of A and B.

Run the program with $N = 20$, $a = 5$, $b = 3$, $\sigma^2 = 1$ and $N1 = 100$. Hence verify (within the limits of this type of investigation) that the least squares estimators of A and B are unbiased, and that their standard errors are as quoted in Section 11.2.

4 Extend the program from question 1 to plot out the observed values of Y at each X, and also (on the same diagram) the values of Y predicted from the fitted regression line.

5† Use Algorithm 11.3, *Multiple Regression*, to examine the relation between W, to be taken as dependent variable, and H, A, taken as independent variables, in the data set of

Appendix C. Carry out the regression separately for males and females, and base each regression on a sample of 20 chosen at random from the whole data set.

6†   Adapt Algorithm 11.2, *Output–Linefit*, to print the output from a multiple regression calculation in the form proposed at the end of Section 11.4.

Use this to complete question 5 by printing out the results for both sets of data.

7†   Continuing questions 5 and 6, carry out regression analyses, using Algorithm 11.1, of (i) W on H, and (ii) W on A, for males and for females separately. Hence examine whether predicting W from both independent variables together, in a multiple regression, gives better results than prediction using only one of the independent variables. (Measure this by comparing the sums of squares for regression with one independent variable and with two independent variables: the difference, with one degree of freedom, is the improvement due to including the second variable.)

12 Analysis of variance

12.1 Introduction

Data collected by means of experiments are usually analysed within the framework of an Analysis of Variance. This technique splits the total variability among all the observations into components. One of these components will correspond to the differences between the effects of the experimental treatments on the experimental units, which may be, for example, plants, animals, or industrial output. Often the units are grouped into blocks, each of which contains relatively uniform units; a component in the analysis will then correspond to differences between these blocks. The final component is the residual variability among units. When carrying out the analysis, the treatment variation is compared with the residual variability, and if it is about the same size we may conclude that the treatments are all having a similar effect; if, however, treatment variation is significantly larger than residual variation we conclude that there are real differences between the effects of the treatments.

The actual quantity which we split into components is the sum of the squares, $\Sigma(x_i - \bar{x})^2$, of the differences between each individual observation x_i and the mean of the whole data set $\bar{x}$. The components are called the **treatment sum of squares**, the **block sum of squares** (if blocks have in fact been used in the experiment) and the **residual sum of squares**. There are certain 'overall' tests which can be made in an analysis of variance, but the precise information on the performance of each experimental treatment is best shown in a table of means. A table of residuals is also useful, to check (as in regression analysis) whether the assumptions underlying the analysis are being satisfied.

In this chapter we present algorithms for analysing data from the two most common types of designed experiment: completely randomised and randomised block designs. Similar forms of analysis of variance are sometimes used with non-experimental data, so we shall name the analysis of variance algorithms according to the structure of the data: there may be **one-way** (single classification) data, as in a completely randomised design, or **two-way** (cross-classification) data as in a randomised block design. A description of the collection and analysis of experimental data, and a discussion of the principles of the analysis of variance, will be found in Clarke (1980); Walpole (1974) also discusses the analysis of variance.

12.2 Input of data

When a set of data arises from a simple designed experiment, each observation can be labelled by its treatment and replicate numbers. In a randomised block design, replicate j of each treatment comes from block j, this collection (block) of units having been chosen to be as similar as possible. With a completely randomised design, however, replicate j of treatment i is not related at all to replicate j of treatment k. The algorithms for both types of analysis of variance operate on a matrix of data $X(\cdot, \cdot)$ classified by treatments (rows) and replicates (columns—which also correspond to blocks in a randomised block design). In a completely randomised design, the number of replicates of each treatment may vary; if it does, some of the cells of the corresponding matrix $X(\cdot, \cdot)$ of data will be empty.

So as to reduce copying errors, it is desirable to input data into a computer directly from experimental record sheets, without any intermediate stage. We therefore recommend that the input program should not require the observations to be entered in any particular order (such as by blocks or by treatments), but instead allow each observation to be identified by its treatment and replicate numbers. Thus an observation might be input using the following instructions:

```
100   INPUT "TREATMENT.";  I
110   INPUT "REPLICATE.";  J
120   INPUT "VALUE......"; X(I,J)
```

Often a number of variates will be recorded at the same time on each experimental unit, e.g. total weight of apples from a tree, weight of grade one apples and total number of apples. It is easy to extend the instructions to input all these data simultaneously, each into its own separate data matrix. We recommend that the data be input by an algorithm such as *Input – Data Array* (4.1) which incorporates adequate checks as the data are entered.

.12.3   One-way analysis of variance

The first part of Algorithm 12.1, *Anova – One Way*, calculates the treatment totals and means, and in the second part these are used to derive the sums of squares. Each **plot residual**, R0, which is the difference between the observation (or 'yield') on that plot and the mean of all plots receiving the same treatment, is calculated and stored. The sum of the squares of these residuals gives the residual sum of squares, which is a component of the analysis of variance table. The treatments sum of squares is calculated from the differences between the treatment means and the overall mean.

```
500   REM   ** ANOVA - ONE WAY **
510   REM   * INPUT: DATA X(.,.), (TREATMENTS * REPLICATES) *
520   REM   * N1 TREATMENTS *
530   REM   * N(I) REPLICATES OF TREATMENT I *
540   REM   * VARIABLES: I, J, M, N, S1, S2, T, T1, T2, W *
550   DIM M(10), R(10,10)

560   REM   CALCULATE MEANS
570   LET N = 0
580   LET T = 0
590       FOR I = 1 TO N1
600       LET T1 = 0
610           FOR J = 1 TO N(I)
620           LET T1 = T1 + X(I,J)            Treatment total
630           NEXT J
640       LET M(I) = T1/N(I)                  Treatment mean
650       LET T = T + T1                      Overall total
660       LET N = N + N(I)                    Total number of observations
670       NEXT I
680   LET M = T/N                             Overall mean
```

```
690   REM   CALCULATE SUMS OF SQUARES
700   LET S1 = 0
710   LET S2 = 0
720       FOR I = 1 TO N1
730       LET T2 = 0
740           FOR J = 1 TO N(I)
750           LET W = X(I,J) - M(I)
760           LET T2 = T2 + W*W
770           LET R(I,J) = W
780           NEXT J
790       LET S2 = S2 + T2
800       LET W = M(I) - M
810       LET S1 = S1 + N(I)*W*W
820       NEXT I

830   REM   * OUTPUT: GRAND MEAN M *
840   REM   * TREATMENT MEANS M(•) *
850   REM   * SUMS OF SQUARES: TREATMENTS S1 *
860   REM   * RESIDUAL S2, PLOT RESIDUALS R(•,•) *
```

SS within treatment I
Plot residual

Residual SS

Treatment SS

Algorithm 12.1 *Anova – One Way*

Algorithm 12.2, *Output – Anova and Means Table (One Way)*, presents the results of the analysis in the following form (in which we have inserted in italics the names of the expressions of the algorithm in the positions that the numerical values would occupy in an actual output).

```
SOURCE        SS      DF      MS

TREATMENTS    S1      K1      S1/K1
RESIDUAL      S2      K2      S2/K2
TOTAL         S1+S2   N-1

F(K1,K2) = F

TREATMENT     MEAN    ST. ERROR

    1         M(1)    SQR(S2/(N(1)*K2))
    2         M(2)    SQR(S2/(N(2)*K2))
    ⋮          ⋮              ⋮
    N1        M(N1)   SQR(S2/(N(N1)*K2))


OVERALL MEAN  M
```

The plot residuals indicate how well the statistical model underlying the analysis of variance fits the data: the smaller the residuals the better the fit. They may be analysed in a similar way to the residuals produced in regression analysis (Section 11.3), in order to check how well the underlying assumptions appear to be satisfied.

Algorithm 12.3, *Output – Table of Residuals (Anova)*, produces a print of observations and corresponding residuals, tabulated by treatment and replicate.

```
500   REM   ** OUTPUT - ANOVA AND MEANS TABLE (ONE WAY) **
510   REM   * INPUT: GRAND MEAN M, NUMBER OF DATA N *
520   REM   * TREATMENT MEANS M(I), NUMBER OF TREATMENTS N1 *
530   REM   * SUMS OF SQUARES: TREATMENTS S1, RESIDUAL S2 *
540   REM   * N(I) REPLICATES OF TREATMENT I *
550   REM   * ROUNDING PARAMETER FØ *
560   REM   * VARIABLES: F, I, K1, K2, W *

570   DEF FNR(X) = INT(X*(1Ø↑FØ) + Ø.5)/(1Ø↑FØ)
580   LET K1 = N1 - 1                                    Treatment df
590   LET K2 = N - N1                                    Residual df

600   REM   ANOVA TABLE
610   PRINT
620   PRINT "SOURCE˄˄˄˄˄˄˄SS˄˄˄˄˄˄˄˄DF˄˄˄˄MS"
630   PRINT
640   PRINT "TREATMENTS"; TAB(11); FNR(S1); TAB(21); K1;
650   PRINT TAB(27); FNR(S1/K1)
660   PRINT "RESIDUAL"; TAB(11); FNR(S2); TAB(21); K2;
670   PRINT TAB(27); FNR(S2/K2)
680   PRINT "TOTAL"; TAB(11); FNR(S1 + S2); TAB(21); N - 1
690   PRINT
700   LET F = (S1/K1)/(S2/K2)
710   PRINT "F("; K1; ","; K2; ") = "; F

720   REM   TABLE OF MEANS
730   PRINT
740   PRINT "TREATMENT˄˄˄MEAN˄˄˄˄˄ST. ERROR"
750   PRINT
760      FOR I = 1 TO N1
770      LET W = SQR(S2/(N(I)*K2))
780      PRINT TAB(6); I; TAB(12); FNR(M(I)); TAB(2Ø); FNR(W)
790      NEXT I
800   PRINT "OVERALL MEAN "; FNR(M)
```

Algorithm 12.2 *Output – Anova and Means Table (One Way)*

```
500      REM   ** OUTPUT - TABLE OF RESIDUALS (ANOVA) **
510      REM   * INPUT: DATA X(•,•), RESIDUALS R(•,•) *
520      REM   * NUMBER OF TREATMENTS N1, N(I) REPLICATES OF TR(I) *
530      REM   * VARIABLES: I, J *

540      FOR I = 1 TO N1
550      PRINT
560      PRINT
570      PRINT "TREATMENT˄"; I
580      PRINT
590      PRINT "REPLICATE˄˄˄OBSERVATION˄˄˄RESIDUAL"
600         FOR J = 1 TO N(I)
610         PRINT TAB(6); J; TAB(12); X(I,J); TAB(26); R(I,J)
620         NEXT J
630      NEXT I
```

Algorithm 12.3 *Output – Table of Residuals (Anova)*

```
500   REM   ** ANOVA - TWO WAY **
510   REM   * INPUT: DATA X(•,•), (TREATMENTS * BLOCKS) *
520   REM   * N1 TREATMENTS, N2 BLOCKS *
530   REM   * VARIABLES: I, J, M, N, S1, S2, S3, T, T1, W *
540   DIM M(20), R(10,10), T(10)

550   REM   CALCULATE MEANS
560       FOR J = 1 TO N2
570       LET T(J) = 0
580       NEXT J
590   LET T = 0
600       FOR I = 1 TO N1
610       LET T1 = 0
620           FOR J = 1 TO N2
630           LET W = X(I,J)
640           LET T = T + W                      Overall sum
650           LET T1 = T1 + W                    Treatment sum
660           LET T(J) = T(J) + W                Block sum
670           NEXT J
680       LET M(I) = T1/N2                       Treatment mean
690       NEXT I
700       FOR J = N1 + 1 TO N1 + N2
710       LET M(J) = T(J - N1)/N1                Block mean
720       NEXT J
730   LET N = N1*N2
740   LET M = T/N                                Overall mean

750   REM   CALCULATE SUMS OF SQUARES
760   LET T = 0
770       FOR I = 1 TO N1
780       LET W = M(I) - M
790       LET T = T + W*W
800       NEXT I
810   LET S1 = N2*T                              Treatment SS
820   LET T = 0
830       FOR J = 1 TO N2
840       LET W = M(N1 + J) - M
850       LET T = T + W*W
860       NEXT J
870   LET S2 = N1*T                              Block SS
880   LET T = 0
890       FOR I = 1 TO N1
900           FOR J = 1 TO N2
910           LET W = X(I,J) - M(I) - M(N1 + J) + M
920           LET R(I,J) = W
930           LET T = T + W*W
940           NEXT J
950       NEXT I
960   LET S3 = T                                 Residual SS

970   REM   * OUTPUT: GRAND MEAN M *
980   REM   * N1 TREATMENT AND N2 BLOCK MEANS IN M(•) *
990   REM   * SUMS OF SQUARES: TREATMENTS S1 *
1000  REM   * BLOCKS S2, RESIDUAL S3 *
1010  REM   * PLOT RESIDUALS R(•,•) *
```

Algorithm 12.4 *Anova – Two Way*

12.4   Two-way analysis of variance

Algorithms 12.4, *Anova–Two Way*, and 12.5, *Output – Anova and Means Table* (*Two Way*), are very similar to the one-way Anova algorithms. The block means are stored in elements N1 + 1 to N1 + N2 of the array M($\cdot$). They are used in the analysis, but are not printed out with the treatment means; it would be easy to have them printed out if

```
500   REM   ** OUTPUT - ANOVA AND MEANS TABLE (TWO WAY) **
510   REM   * INPUT: GRAND MEAN M *
520   REM   * TREATMENT MEANS M(•), ROUNDING PARAMETER FØ *
530   REM   * SUMS OF SQUARES: TREATMENTS S1 *
540   REM   * BLOCKS S2, RESIDUAL S3 *
550   REM   * NUMBER OF TREATMENTS N1, OF BLOCKS N2 *
560   REM   * VARIABLES: F1, F2, I, K1, K2, K3 *

570   DEF FNR(X) = INT(X*(10↑FØ) + Ø.5)/(10↑FØ)
580   LET K1 = N1 - 1                                    Treatment df
590   LET K2 = N2 - 1                                    Blocks df
600   LET K3 = N1*N2 - N1 - N2 + 1                       Residual df

610   REM   ANOVA TABLE
620   PRINT
630   PRINT "SOURCE^^^^^SS^^^^^^^^^DF^^^MS"
640   PRINT
650   PRINT "TREATMENTS"; TAB(11); FNR(S1); TAB(21); K1;
660   PRINT TAB(27); FNR(S1/K1)
670   PRINT "BLOCKS"; TAB(11); FNR(S2); TAB(21); K2;
680   PRINT TAB(27); FNR(S2/K2)
690   PRINT "RESIDUAL"; TAB(11); FNR(S3); TAB(21); K3;
700   PRINT TAB(27); FNR(S3/K3)
710   PRINT "TOTAL"; TAB(11); FNR(S1 + S2 + S3); TAB(21); N1*N2
720   LET F1 = (S1/K1)/(S3/K3)
730   LET F2 = (S2/K2)/(S3/K3)
740   PRINT
750   PRINT "TREATMENTS F("; K1; ","; K3; ") = "; F1
760   PRINT "BLOCKS^^^^^^F("; K2; ","; K3; ") = "; F2
770   PRINT

780   REM   TABLE OF MEANS
790   PRINT
800   PRINT "TREATMENT^^^MEAN"
810   PRINT
820       FOR I = 1 TO N1
830       PRINT TAB(6); I; TAB(12); FNR(M(I))
840       NEXT I
850   PRINT "OVERALL MEAN "; FNR(M)
860   PRINT "S.E. OF TREATMENT MEAN "; SQR(S3/(K3*N2))
```

Algorithm 12.5 *Output – Anova and Means Table (Two Way)*

needed. The output takes the following form, where the expressions are shown in italics in the positions that their numerical values would occupy in the actual output.

```
SOURCE        SS           DF           MS

TREATMENTS    S1           K1           S1/K1
BLOCKS        S2           K2           S2/K2
RESIDUAL      S3           K3           S3/K3
TOTAL         S1+S2+S3     N1*N2-1

TREATMENTS  F(K1,K3)  =  F1
BLOCKS      F(K2,K3)  =  F2


TREATMENT     MEAN

    1         M(1)
    2         M(2)
    .          .
    .          .
    .          .
   N1         M(N1)



OVERALL MEAN M
S.E OF TREATMENT MEAN SQR(S3/(K3*N2))
```

Algorithm 12.3, *Output – Table of Residuals (Anova)*, may be used for the two-way analysis of variance. When the input variables of the algorithm are being set, all the elements of N(·) should be put equal to N2.

12.5   Test data for Algorithms

Algorithm 12.1, *Anova – One Way*. Input: N1 = 3, N(·) = (3, 5, 6). X(·, ·) has 1st row 1, 4, 4; 2nd row 3, 6, 7, 9, 10; 3rd row 6, 8, 10, 11, 12, 13. Output: M = 7.43, M(·) = (3, 7, 10), S1 = 99.43, S2 = 70. R(·, ·) has 1st row $-2, 1, 1$; 2nd row $-4, -1, 0, 2, 3$; 3rd row -4, $-2, 0, 1, 2, 3$.

Algorithm 12.4, *Anova – Two Way*. Input: N1 = 3, N2 = 4. X(·, ·) has 1st row 1, 2, 3, 6; 2nd row 2, 6, 7, 9; 3rd row 8, 10, 11, 15. Output: M = 6.6̇6̇, M(·) = (3, 6, 11, 3.66, 6, 7, 10), S1 = 130.67, S2 = 62, S3 = 4. R(·, ·) has 1st row 1, $-\frac{1}{3}$, $-\frac{1}{3}$, $-\frac{1}{3}$; 2nd row $-1, \frac{2}{3}, \frac{2}{3}$, $-\frac{1}{3}$; 3rd row 0, $-\frac{1}{3}$, $-\frac{1}{3}, \frac{2}{3}$.

12.6   Exercises

1   Write an input algorithm for experimental data, on the lines suggested in Section 12.3, capable of handling several variates at one time.

2   Construct a program which analyses data from a completely randomised design, and which
  i) inputs experimental data using a checking routine as suggested in Section 12.2;
  ii) carries out a one-way analysis of variance on the data (Algorithm 12.1);

iii) prints out the analysis of variance table and a table of means (Algorithm 12.2).
Use this complete program with the following data.

Treatment	Replicate	Value	Treatment	Replicate	Value
1	1	24	2	3	37
2	1	46	3	4	41
1	2	18	2	4	50
1	3	18	2	5	44
3	1	32	3	5	36
3	2	30	3	6	28
1	4	29	1	7	15
2	2	39	2	6	45
1	5	22	2	7	30
1	6	17	3	7	27
3	3	26			

3   Construct a program, similar to that described in question 2, but for a two-way analysis of variance using Algorithms 12.4 and 12.5.

Use this program with the following data.

Treatment	Block	Value	Treatment	Block	Value
1	1	68	4	3	103
2	1	71	1	3	77
3	1	54	5	3	88
2	2	78	2	3	74
3	2	67	3	3	65
4	1	95	1	4	59
5	1	73	2	4	70
1	2	82	4	4	90
4	2	116	3	4	54
5	2	85	5	4	76

4   Add a print-out of residuals, using Algorithm 12.3, to each of the programs written in questions 2 and 3.

5   (a) The residual in an observation from a completely randomised experiment is the difference between the observation and the mean of all observations on the same treatment. Use Algorithm 5.5, *Output – Scattergram*, to plot the residuals against the treatment means (see Section 11.3).

(b) The residual in an observation from a randomised block experiment is $r_{ij} = x_{ij} - m - t_i - b_j$; m is the mean of all observations in the experiment, $m + t_i$ is the mean for treatment i, and $m + b_j$ is the mean for block j. The corresponding fitted value is $m + t_i + b_j$. Plot the residuals against the fitted values.

6 Write a program to:
i) generate N observations according to the model
$$x_{ij} = m + t_i + e_{ij},$$

where $t_i = 4i$ and e_{ij} is a random sample from a normal distribution with mean 0 and variance σ^2; the values taken by j are 1 to n_i, where the $\{n_i\}$ need not all be the same (but must add to N);
ii) carry out a one-way analysis of variance on the observations generated;
iii) compare the treatment means and the variance estimated from the analysis with those used in the generation;
iv) repeat this N9 times and study the distribution of the estimates of variance.

Run the program with $i = 1$ to 4, $n_1 = 7$, $n_2 = 4$, $n_3 = 5$, $n_4 = 6$, $\sigma^2 = 3.24$, $m = 1$, N9 = 50.

7 Write a program to
i) generate N observations according to the model
$$x_{ij} = m + t_i + b_j + e_{ij},$$

where $t_i = 3i$ and $b_j = j + 5$, and e_{ij} is a random sample from a normal distribution with mean 0 and variance σ^2; the values of i run from 1 to t_0 and of j from 1 to r_0, where $r_0 t_0 = N$;
ii) carry out a two-way analysis of variance on the observations generated;
iii) compare the treatment means and the variance estimated from the analysis with those used in the generation;
iv) repeat this N1 times and study the distribution of the estimates of variance.

Run the program with $i = 6$, $j = 4$, $m = 5$, $\sigma^2 = 2.25$, N1 = 50.

8 It is sometimes useful to analyse a transformed function of the observations. Amend the program for one-way analysis of variance so that it can also operate on the logarithms of the original observations, and print out an appropriate message when it has done so.

Use the amended program to analyse the following experimental data, carrying out a log transformation first.
Treatment A: 12.1, 16.3, 11.8, 14.5, 13.7, 15.2
Treatment B: 17.5, 25.4, 22.6, 31.3, 42.0, 28.0, 37.5, 34.8
Treatment C: 50.3, 39.1, 47.4, 62.6, 48.3, 57.7, 35.9

9† Use Algorithm 12.1, *Anova – One Way*, to compare the mean of H (Student Data Set, Appendix C) for males with its mean for females. Repeat for W, R, and C of Appendix C.

Appendix A Using the Algorithms in this book

A.1 Description of Algorithms

The form of the algorithms may be illustrated by a simple one which calculates a mean (see Example A.1).

```
500   REM   ** MEAN OF ARRAY **
510   REM   * INPUT: DATA X(•), NUMBER OF DATA N *
520   REM   * VARIABLES: I, T *

530   LET T = 0
540       FOR I = 1 TO N
550       LET T = T + X(I)
560       NEXT I
570   LET M = T/N

580   REM   * OUTPUT: MEAN M *
```

Example A.1

The following rules have been used when writing the algorithms in the body of this book. (The algorithms in Appendix B have been written in a compact form and, in these, Rules 2, 3, 13 and 14 are not followed.)

1) The *subset of BASIC* used is that described in the appendix of *Basic BASIC* (Monro, 1978).
2) There is only *one statement* on each line.
3) The LET statement is always used when assigning a value to a variable.
4) *Line numbers* begin at 500, and increase by units of 10. (*Programs* begin at line 10.)
5) Each algorithm is given a *title* inserted between double asterisks as part of a REM statement, e.g. 500 REM ** MEAN OF ARRAY **.
6) The algorithms containing input or output statements are given a title beginning with INPUT or OUTPUT, whichever is appropriate; no algorithm contains both input and output statements. All the other algorithms, not so titled, contain neither input nor output statements.
7) The variables required for *input* to an algorithm are listed immediately after the title. Each variable is briefly described and the list is inserted between single asterisks, in a REM statement, and introduced by the word INPUT, e.g.
   ```
   510   REM   * INPUT: DATA X(•), NUMBER OF DATA N *
   ```
   The list may continue over several lines.
8) If an algorithm makes use of a *subroutine*, the name of the subroutine is given on a separate line and inserted between single asterisks in a REM statement, and introduced by the word REQUIRES. The line number which is given in the algorithm for the start of the subroutine is also stated, e.g.
   ```
   520   REM   * REQUIRES: MAXMIN (3.5) AT 1500 *
   ```

9) The *variables* which are introduced in an algorithm, apart from those listed in the input and those specified in a DIM statement (see Rule 10) are listed between single asterisks, in a REM statement, and introduced by the word VARIABLES, e.g.

```
52Ø    REM   * VARIABLES: I, T *
```

10) *Dimension* statements for *arrays* introduced in an algorithm follow immediately after the initial descriptive lines.

11) The major variables calculated in an algorithm are listed in the last line, or last few lines, of the algorithm. The list is inserted between single asterisks, in a REM statement, and is introduced by the word OUTPUT, e.g.

```
58Ø    REM   * OUTPUT: MEAN M *
```

Note that if an algorithm is used as a subroutine the final line is required to be a RETURN statement.

12) REM statements, without asterisks, are used to label blocks of statements in large algorithms, to describe subroutines when they are called, and to point out special things to note.

13) Blocks of statements are separated by *spaces*. Unfortunately these are lost, in most BASICs, when the algorithm is reproduced by the computer; however, forms of separation that are preserved are

```
6ØØ    REM   *********************
7ØØ    REM
```

14) FOR—NEXT loops are indented. This indentation also will be lost when the algorithm is reproduced by a computer.

A.2   Running an Algorithm

The algorithms may be classified as (1) input, (2) output, or (3) calculation, algorithms. Only calculation algorithms are considered in this section. The simplest way to run one of these algorithms is to put input instructions before it and output instructions after it. The descriptive part of the algorithm lists the required input variables and the output variables produced by the algorithm (see Example A.2).

```
1Ø    REM   ** TITLE **
2Ø    DIM X(1Ø)

3Ø    INPUT N
4Ø        FOR I = 1 TO N
5Ø        INPUT X(I)
6Ø        NEXT I

5ØØ   REM   ** MEAN OF ARRAY **
51Ø   REM   * INPUT: DATA X(•), NUMBER OF DATA N *
•••   (Algorithm, Example A.1)
58Ø   REM   * OUTPUT: MEAN M *

59Ø   PRINT M
6ØØ   END
```

Example A.2

A command RUN will run the program; the data are entered in response to the requests for input by the computer. (This type of input is adequate when checking algorithms but, as we argue in Chapter 2, for serious data analysis it is better to input by a READ statement or an input algorithm that incorporates checks.)

An alternative method is to use the algorithm as a subroutine. One part of the program, which might be called the **control section**, prepares variables for input into the algorithm, which is called as a subroutine, and prints out results after program execution returns to the control section from the algorithm (subroutine). Note that, when an algorithm is used in this way, a RETURN statement must be added at the end of it. Example A.3 serves as an illustration.

```
10   REM  ** TITLE **
20   DIM X(10)

30   INPUT N
40       FOR I = 1 TO N
50       INPUT X(I)
60       NEXT I

70   REM  CALCULATE MEAN
80   GOSUB 500
90   PRINT M
100  END

500  REM  ** MEAN OF ARRAY **
...  (Algorithm, Example A.1)
580  REM  * OUTPUT: MEAN M *
590  RETURN
```

Example A.3

A.3   Combining Algorithms

Suppose we wish to find the mean and the maximum and minimum values of a set of data. We require two algorithms: *Mean of Array* (Example A.1) and *Maxmin* (Algorithm 3.5). There are two main methods of combination: (1) chaining; (2) using subroutines called from a control section.

A.3.1   Chaining Algorithms

The method is simply to follow one algorithm by another. Additional instructions may be needed for input and output (although specific input and output algorithms may be introduced), to prepare variables for input to another algorithm, and to control the progress through the program. Example A.4, a program to solve the problem given at the beginning of this section, is a simple illustration.

```
10   REM  ** TITLE **
20   DIM X(10)

30   INPUT N
40       FOR I = 1 TO N
50       INPUT X(I)
60       NEXT I
```

```
5ØØ   REM   ** MEAN OF ARRAY **
51Ø   REM   * INPUT: DATA X(•), NUMBER OF DATA N *
52Ø   REM   * VARIABLES: I, T *
•••   (Algorithm, Example A.1)
58Ø   REM   * OUTPUT: MEAN M *

15ØØ   REM   ** MAXMIN **
151Ø   REM   * INPUT: DATA X(•), NUMBER OF DATA N *
152Ø   REM   * VARIABLES: I, L, U, W *
••••   (Algorithm 3.5)
163Ø   REM   * OUTPUT: MAXIMUM U, MINIMUM L *

164Ø   PRINT M, U, L
165Ø   END
```

Example A.4

A.3.2 Subroutines

Example A.3 is easily extended to include an additional subroutine which calculates maximum and minimum values. Algorithm 3.5, *Maxmin*, might be placed at lines 1500–1650 (including a RETURN at the end). The addition of the instruction

```
85    GOSUB 15ØØ
```

together with the extension of line 90 to

```
9Ø    PRINT M, U, L
```

completes the program.

In more complex examples more instructions will be needed in the control section to set up appropriate variables for input into the subroutines, or to regulate the flow of execution. In the example, all input and output takes place in the control section. With more realistic examples of data analysis, both input and output will probably be done with appropriate algorithms which themselves might be called from the control section.

A convenience of this approach is that as each algorithm is added, as a subroutine, the control section can be rewritten so that a test may be made of the new subroutine. In this way a complex program is built up in which each part has been properly tested.

A.4 The menu approach

This is a method, which may be used with an interactive system, for controlling the flow of execution between algorithms. It may be used with chained algorithms or with subroutines. The approach is particularly convenient when analysing data, because it allows the operator to choose which algorithm is executed next, depending on how the analysis is proceeding. The essence of the method is that a list of algorithms, or **menu**, is offered to the operator who chooses which algorithm is obeyed next, after which the program returns to the menu. The operator may choose another algorithm or stop the analysis.

An outline of a program which offers a choice of the algorithms *Mean of Array* (Example A.1) or *Maxmin* (3.5), using a control section and subroutines, is shown in Example A.5.

```
10   REM  ** TITLE **
20   DIM X(10)

30   INPUT N
40      FOR I = 1 TO N
50      INPUT X(I)
60      NEXT I

70   PRINT "1. MEAN OF ARRAY"
80   PRINT
90   PRINT "2. MAXIMUM AND MINIMUM"
100  PRINT
110  PRINT "TYPE NUMBER OF ITEM REQUIRED AND PRESS RETURN"
120  INPUT I

130  IF I <> 1 THEN 170
140  GOSUB 500
150  PRINT "MEAN ": M
160  GO TO 70

170  IF I <> 2 THEN 220
180  GOSUB 1500
190  PRINT "MAXIMUM "; U
200  PRINT "MINIMUM "; L
210  GO TO 70

220  PRINT "NUMBER NOT RECOGNISED - REPEAT"
230  GO TO 70
240  END

500  REM  ** MEAN OF ARRAY **
...  (Algorithm, Example A.1)
590  RETURN

1500 REM  ** MAXMIN **
.... (Algorithm 3.5)
1640 RETURN
```

Example A.5

Program 6.6, *Bivariate Sample Analysis*, is an example of a menu program in which the algorithms are chained (see page 60).

A.5   Points to note when combining Algorithms

(1)   *Line numbers.* Since the line numbers of all the algorithms begin at 500, the line numbers of at least one algorithm will need changing when algorithms are combined. In the examples we give, we have added multiples of 1000 to line numbers, which makes the change simple; other methods may be adopted if this is thought to string out the line numbers too much. Note that the line numbers in jump instructions in the body of the algorithm will also need changing.

(2)   *Variables.* Check whether the same variables are used in different algorithms. (A list of the variables used in each algorithm is given at the head of the algorithm.) This is

allowable if variables are re-set and used in distinct loops (e.g. counters, and variables which sum), but do *not* use a subroutine, which contains a loop with counter I (say), in the *middle* of a loop in the control section which also uses the same counter variable. In Example A.4 we see that the two algorithms both include a variable I, but the loops controlled by I are quite distinct in the two algorithms, and therefore in this case the duplication of variable in the full program is allowable.

(3) *Dimensions*. Make sure that all arrays are dimensioned appropriately, so that they are at least as large as the program requires. Each algorithm in this book includes dimension statements for the arrays introduced in the algorithm. All the dimension statements may be put at the beginning of a program, if this is preferred. In some versions of BASIC the dimensions of an array may be changed as the program proceeds.

Appendix B Compact Algorithms

In this Appendix, more than one program instruction appears on the same line. Colons (:) are used as separators, and 'LET' is omitted.

Algorithm A.1 *Quicksort*

```
500    REM  ** QUICKSORT **
510    REM  * INPUT: DATA ARRAY A(•), NUMBER OF DATA N *
520    REM  * REQUIRES: EXCHANGE SORT AT 1000 *
530    REM  * VARIABLES: D, F, I, J, K, K1, L, R, W *
540    DIM H(10)

550    D = 10 : K = 1 : L = 1 : R = N
560    REM  EXCHANGE SORT IF BLOCKSIZE <= D
570    IF R - L + 1 > D THEN 590
580    GOSUB 1000 : GO TO 780
590    I = L : J = R : W = RND(1)*(R - L) + 0.5
600    W = INT(W + L) : F = A(W)

610    REM  PARTITION PHASE
620    IF A(I) >= F THEN 640
630    I = I + 1 : GO TO 620
640    IF A(J) <= F THEN 660
650    J = J - 1 : GO TO 640
660    IF I > J THEN 680
670    W = A(I) : A(I) = A(J) : A(J) = W : I = I + 1 : J = J - 1
680    IF I <= J THEN 620

690    REM  BOOKKEEPING PHASE
700    IF J - L >= R - I THEN 740
710    IF I >= R THEN 730
720    H(K) = I : K = K + 1 : H(K) = R : K = K + 1
730    R = J : GO TO 770
740    IF L >= J THEN 760
750    H(K) = L : K = K + 1 : H(K) = J : K = K + 1
760    L = I
770    IF L < R THEN 590
780    IF K <= 1 THEN 800
790    K = K - 1 : R = H(K) : K = K - 1 : L = H(K) : GO TO 560
800    REM  * OUTPUT: DATA IN ASCENDING ORDER IN A(•) *

1000   REM  ** EXCHANGE SORT **
1010   FOR I = L TO R - 1 : K1 = I
1020   FOR J = I + 1 TO R : IF A(J) >= A(K1) THEN 1040
1030   K1 = J
1040   NEXT J
1050   IF K1 = I THEN 1070
1060   W = A(I) : A(I) = A(K1) : A(K1) = W
1070   NEXT I : RETURN
```

Algorithm A.2   *Monkey Puzzle Sort*

```
500   REM  ** MONKEY PUZZLE SORT **
510   REM  * INPUT: DATA A(•), NUMBER OF DATA N *
520   REM  * VARIABLES: I, J, K *
530   DIM B(100), L(100), R(100)

540   REM  CONSTRUCT TREE
550   L(1) = 0 : R(1) = 0
560   FOR I = 2 TO N
570   L(I) = 0 : R(I) = 0 : J = 1
580   IF A(I) > A(J) THEN 620
590   IF L(J) = 0 THEN 610
600   J = L(J) : GO TO 580
610   R(I) = -J : L(J) = I : GO TO 650
620   IF R(J) <= 0 THEN 640
630   J = R(J) : GO TO 580
640   R(I) = R(J) : R(J) = I
650   NEXT I

660   REM  STRIP THE TREE
670   J = 1 : K = 1 : GO TO 690
680   J = L(J)
690   IF L(J) > 0 THEN 680
700   B(K) = A(J) : K = K + 1
710   IF R(J) = 0 THEN 750
720   IF R(J) < 0 THEN 740
730   J = R(J) : GO TO 690
740   J = -R(J) : GO TO 700

750   REM  * OUTPUT: DATA IN ASCENDING ORDER IN B(•) *
```

Algorithm A.3   *F Distribution Fnc*

```
500   REM   ** F DISTRIBUTION FNC **
510   REM   * INPUT: ARGUMENT F, DEGREES OF FREEDOM K1, K2 *
520   REM   * ACCURACY E *
530   REM   * REQUIRES: LOG GAMMA (8.6) AT 1500 *
540   REM   * VARIABLES: G1,G2,H1,H2,H3,I,PØ,T1,T2,W,W1,W2,W3,W4 *

550   H2 = K2/2 : H1 = K1/2 : H3 = H1 + H2
560   W = H2 : GOSUB 1500 : G2 = G1
570   W = H1 : GOSUB 1500 : G2 = G2 + G1
580   W = H3 : GOSUB 1500 : G2 = G2 - G1
590   W1 = K2/(K2 + K1*F) : W = Ø
600   IF H2 >= H3*W1 THEN 620
610   W2 = W1 : W1 = 1 - W1 : W3 = H2 : H2 = H1 : H1 = W3 : GO TO 630
620   W2 = 1 - W1 : W = 1
630   T1 = 1 : W3 = 1 : PØ = 1 : I = INT(H1 + W2*H3) : W4 = W1/W2
640   T2 = H1 - W3 : IF I <> Ø THEN 660
650   W4 = W1
660   T1 = T1*T2*W4/(H2 + W3) : PØ = PØ + T1 : T2 = ABS(T1)
670   IF T2 <= E AND T2 <= E*PØ THEN 700
680   W3 = W3 + 1 : I = I - 1 : IF I >= Ø THEN 640
690   T2 = H3 : H3 = H3 + 1 : GO TO 660
700   W1 = EXP(H2*LOG(W1) + (H1 - 1)*LOG(W2) - G2)
710   PØ = PØ*W1/H2 : IF W = Ø THEN 730
720   PØ = 1 - PØ

730   REM   * OUTPUT: PØ, PROBABILITY X < F *
```

$$\ln\left(\frac{\Gamma(h_1)\Gamma(h_2)}{\Gamma(h_1 + h_2)}\right)$$

Algorithm A.4 *Inverse F*

```
500   REM   ** INVERSE F **
510   REM   * INPUT: PROBABILITY PØ, DEGREES OF FREEDOM K1, K2 *
520   REM   * ACCURACY PARAMETER E *
530   REM   * REQUIRES: F DISTRIBUTION FNC AT 1ØØØ *
540   REM   * LOG GAMMA AT 15ØØ *
550   REM   * VARIABLES: F, G1, G2, G3, H1, H2, H3, J1, J2 *
560   REM   * P1, W, W1, W2, X, X1 *

570   H1 = K2/2 : H2 = K1/2 : H3 = H1 + H2
580   W = H1 : GOSUB 15ØØ : G2 = G1
590   W = H2 : GOSUB 15ØØ : G2 = G2 + G1
600   W = H3 : GOSUB 15ØØ : G2 = G2 - G1
610   P1 = 1 - PØ : J1 = H1 : J2 = H2
```

$$\ln\left(\frac{\Gamma(h_1)\Gamma(h_2)}{\Gamma(h_1+h_2)}\right)$$

```
620   REM   CALCULATE INITIAL ESTIMATE FOR X
630   W = SQR(-LOG(P1*P1))
640   W = W - (2.3Ø753 + Ø.27Ø61*W)/(1 + W*(Ø.99229 + Ø.Ø4481*W))
650   W1 = 2*J2 : W2 = 1/(9*J2)
660   W2 = W1*(1 - W2 + W*SQR(W2))↑3 : IF W2 <= Ø THEN 69Ø
670   W2 = (4*J1 + W1 - 2)/W2 : IF W2 <= 1 THEN 7ØØ
680   X = 1 - 2/(1 + W2) : GO TO 71Ø
690   X = 1 - EXP((LOG((1 - P1)*J2) + G2)/J2) : GO TO 71Ø
700   X = EXP((LOG(P1*J1) + G2)/J1)
710   W1 = 1 - J1 : W2 = 1 - J2

720   REM   NEWTON ITERATION
730   GOSUB 1ØØØ
740   W = (P2 - P1)*EXP(W1*LOG(X) + W2*LOG(1 - X) + G2)
750   X1 = X - W : IF X1 < 1.Ø AND X1 > Ø THEN 77Ø
760   W = W/2 : GO TO 75Ø
770   X = X1 : IF ABS(W) > E THEN 73Ø
780   F = J1*(1 - X)/(J2*X)
790   REM   * OUTPUT: F-VALUE F *

1000  REM   ** DISTRIBUTION FNC - CALCULATES PROBABILITY P2(X) **
1010  X2 = 1 - X : H4 = H3 : IF J1 >= H3*X THEN 1Ø3Ø
1020  X1 = X2 : X2 = X : J3 = J2 : J4 = J1 : J$ = "T" : GO TO 1Ø4Ø
1030  X1 = X : J3 = J1 : J4 = J2 : J$ = "F"
1040  T1 = 1 : W3 = 1 : P2 = 1 : I = INT(J4 + X2*H4) : W = X1/X2
1050  T2 = J4 - W3 : IF I <> Ø THEN 1Ø7Ø
1060  W = X1
1070  T1 = T1*T2*W/(J3 + W3) : P2 = P2 + T1
1080  T2 = ABS(T1) : IF T2 <= E AND T2 <= E*P2 THEN 111Ø
1090  W3 = W3 + 1 : I = I - 1 : IF I >= Ø THEN 1Ø5Ø
1100  T2 = H4 : H4 = H4 + 1 : GO TO 1Ø7Ø
1110  P2 = P2*EXP(J3*LOG(X1) + (J4 - 1)*LOG(X2) - G2)/J3
1120  IF J$ = "F" THEN 114Ø
1130  P2 = 1 - P2
1140  RETURN

1500  REM   ** LOG GAMMA FUNCTION **
1510  G1 = Ø
1520  W = W - 1 : IF W <= Ø THEN 154Ø
1530  G1 = G1 + LOG(W) : GO TO 152Ø
1540  IF W = Ø THEN 156Ø
1550  G1 = G1 + Ø.57236494
1560  RETURN
```

Algorithm A.5 *Multivar SS and SP (Deviation Method)*

```
500   REM  ** MULTIVAR SS AND SP (DEVIATION METHOD) **
510   REM  * INPUT: DATA X(•,•); N UNITS * C1 VARIABLES *
520   REM  * VARIABLES: I, J, K, T, T1, T2 *
530   DIM M(10), Q(10,10)

540   FOR J = 1 TO C1 : T = 0
550   FOR I = 1 TO N : T = T + X(I,J) : NEXT I
560   M(J) = T/N : NEXT J
570   FOR I = 1 TO C1 : T1 = M(I)
580   FOR J = 1 TO I : T = 0 : T2 = M(J)
590   FOR K = 1 TO N : T = T + (X(K,I) - T1)*(X(K,J) - T2) : NEXT K
600   Q(I,J) = T : NEXT J : NEXT I

610   REM  * OUTPUT: MEANS M(•), SS AND SP MATRIX Q(•,•) *
```

Algorithm A.6   *Linear Solver*

```
1500   REM  ** LINEAR SOLVER **
1510   REM  * INPUT: NUMBER OF EQUATIONS C, C1 = C + 1 *
1520   REM  * LOWER TRIANGLE OF (C + 1)*C COEFF MATRIX IN S(·) *
1530   REM  * (C*C COEFF MATRIX MUST BE SYMMETRIC, PD) *
1540   REM  * ACCURACY PARAMETER E *
1550   REM  * VARIABLES: I,I1,I2,I3,J,J1,J2,J3,N1,N2,N3,S,W *
1560   DIM B(10), W(10)

1570   REM  FACTORISATION
1580   N1 = 0 : J1 = 1 : J2 = 0
1590   FOR J = 1 TO C : I1 = 0
1600   FOR I = 1 TO J : J2 = J2 + 1 : W = S(J2) : I2 = J1
1610   FOR I3 = 1 TO I : I1 = I1 + 1 : IF I3 = I THEN 1630
1620   W = W - S(I1)*S(I2) : I2 = I2 + 1 : NEXT I3
1630   IF I = J THEN 1680
1640   IF S(I1) = 0 THEN 1660
1650   S(J2) = W/S(I1) : GO TO 1670
1660   S(J2) = 0
1670   NEXT I
1680   IF W < E THEN 1700
1690   S(J2) = SQR(W) : GO TO 1720
1700   S(J2) = 0 : N1 = N1 + 1
1710   PRINT "ZERO DIAGONAL ELEMENT S("; J2; ")"
1720   J1 = J1 + J : NEXT J

1730   REM  INVERT S ONTO ITSELF
1740   N2 = C*C1/2 : I = C : N3 = N2
1750   IF S(N3) = 0 THEN 1880
1760   I1 = N3
1770   FOR I3 = I TO C : W(I3) = S(I1) : I1 = I1 + I3 : NEXT I3
1780   J = C : J3 = N2 : I2 = N2
1790   I1 = J3 : W = 0 : IF I <> J THEN 1810
1800   W = 1/W(I)
1810   J2 = C
1820   IF J2 = I THEN 1860
1830   W = W - W(J2)*S(I1) : J2 = J2 - 1 : I1 = I1 - 1
1840   IF I1 <= I2 THEN 1820
1850   I1 = I1 - J2 + 1 : GO TO 1820
1860   S(I1) = W/W(I) : IF J = I THEN 1900
1870   I2 = I2 - J : J = J - 1 : J3 = J3 - 1 : GO TO 1790
1880   I1 = N3
1890   FOR J1 = I TO C : S(I1) = 0 : I1 = I1 + J1 : NEXT J1
1900   N3 = N3 - I : I = I - 1 : IF I <> 0 THEN 1750

1910   REM  FORM SOLUTION VECTOR B(·)
1920   FOR I = 1 TO C : W(I) = S(N2 + I) : NEXT I
1930   I1 = 0 : FOR I = 1 TO C : I1 = I1 + I - 1 : S = 0
1940   FOR J = 1 TO I : I2 = I1 + J : S = S + S(I2)*W(J) : NEXT J
1950   IF I = C THEN 1970
1960   FOR J = I + 1 TO C : I2 = I2 + J - 1 : S = S + S(I2)*W(J) : NEXT J
1970   B(I) = S : NEXT I : RETURN

1980   REM  * OUTPUT: REGRESSION COEFFICIENTS B(·), INV(S) *
1990   REM  * NULLITY FACTOR N1 *
```

We recommend that, in Algorithm A.6, the accuracy parameter E be given a value of 1.0E–6 on microcomputers; on mainframe computers it may be given a much smaller value. If, in the factorisation, a diagonal element is found to be less than E then it, and elements in the same row, will be set to zero. A message "ZERO DIAGONAL ELEMENT", with the variable number, is printed and the nullity factor Nl is increased by 1. If, with highly collinear data, all regression coefficients are printed out as zeros then a change of the accuracy parameter may lead to more sensible results.

Appendix C Student data set

We give here a large data set, collected from 100 university students. The first column is simply the student reference number; the next four columns are self-explanatory (but note that 'years' is quoted as a decimal, not as years and months). The variable 'left-handedness' was recorded on an integer scale from 0 to 20, a high score indicating a very left-handed person based on answers to a questionnaire which listed various activities (e.g. writing, throwing, striking a match) that can be done with either hand. 'Reaction time' was the length of time that elapsed between a light being flashed and the person pressing a button. 'Card sorting time' was the time taken to sort a thoroughly shuffled standard pack of playing cards into the four suits.

References in the text have already suggested uses for parts of these data, and further suggestions follow the data (page 149).

C.1 Student data set

S Sex, M or F
H Height in centimetres
W Weight in kilograms
A Age in years
L Lefthandedness score, 0–20
R Reaction time in milliseconds
C Card sorting time in seconds

Number	S	H	W	A	L	R	C
1	M	176	72	22.6	0	212	33
2	M	172	58	18.5	20	163	56
3	M	177	60	19.1	3	220	50
4	M	174	68	19.6	0	169	66
5	M	167	59	18.9	8	140	62
6	M	191	72	19.9	7	156	63
7	M	190	85	19.4	3	178	95
8	M	176	73	19.0	6	136	52
9	F	165	68	18.2	4	205	28
10	F	167	63	18.0	3	194	48
11	F	172	60	18.5	6	186	51
12	F	169	58	19.6	3	172	63
13	M	173	57	18.9	1	149	71
14	M	180	76	19.2	2	171	57
15	F	154	49	19.9	6	198	59
16	M	178	69	20.1	5	182	72

Number	S	H	W	A	L	R	C
17	M	170	74	19.8	8	257	60
18	M	173	63	19.0	3	155	83
19	M	168	50	21.1	1	139	62
20	M	173	64	20.9	3	137	73
21	M	178	71	19.0	2	162	34
22	F	170	63	19.2	17	176	64
23	F	175	58	18.9	5	175	62
24	F	156	57	20.6	7	207	98
25	F	170	55	18.3	4	169	60
26	M	182	66	18.4	3	226	55
27	M	175	67	18.9	3	171	50
28	F	173	57	19.4	4	188	59
29	M	175	65	20.0	11	177	58
30	M	173	61	19.5	5	148	55
31	M	178	68	19.7	18	131	64
32	M	174	64	19.3	4	263	65
33	M	176	66	29.1	6	159	59
34	F	177	65	22.9	6	156	49
35	F	161	47	20.1	15	193	60
36	M	167	64	18.5	4	140	62
37	F	168	59	33.9	3	162	55
38	M	185	76	34.2	18	167	58
39	F	164	67	24.1	7	254	72
40	F	160	55	29.4	1	277	51
41	M	174	55	18.5	5	172	51
42	F	154	52	18.5	6	231	55
43	M	170	60	20.4	5	186	58
44	M	176	68	19.1	3	174	56
45	M	160	53	22.3	14	176	52
46	M	174	71	28.3	18	153	66
47	F	148	33	19.9	4	172	62
48	F	161	53	25.2	4	205	55
49	M	182	72	28.0	4	210	58
50	M	184	80	19.2	8	201	60
51	F	158	53	19.0	6	187	55
52	F	170	54	18.5	1	162	62
53	F	157	50	19.8	0	145	61
54	F	173	52	18.8	10	95	53
55	F	162	70	18.2	8	207	61
56	F	163	53	19.9	7	159	57
57	M	183	73	17.9	0	220	54
58	M	179	68	19.0	0	220	62
59	M	164	63	23.6	6	160	51
60	M	177	72	18.8	15	260	60
61	M	179	66	23.1	3	160	64
62	F	172	64	18.4	6	240	55
63	F	174	60	18.7	3	170	52
64	F	173	68	18.9	4	175	56
65	F	164	57	18.9	3	245	56
66	M	184	74	20.1	0	275	63
67	M	181	64	20.0	7	200	54
68	F	164	52	19.0	4	176	67

Number	S	H	W	A	L	R	C
69	M	177	67	27.0	0	210	54
70	M	177	69	22.9	8	193	53
71	M	178	73	18.2	5	162	59
72	M	178	71	18.9	9	280	62
73	M	181	60	19.1	9	220	63
74	F	157	49	19.9	19	160	55
75	M	182	70	18.9	8	180	61
76	M	183	63	18.9	0	140	60
77	M	184	92	24.9	4	155	78
78	F	170	57	19.0	15	350	50
79	F	168	54	18.5	2	375	57
80	F	159	53	18.4	3	340	65
81	F	157	49	19.0	5	205	69
82	F	160	50	18.4	7	175	60
83	M	176	61	18.0	3	191	62
84	F	162	52	18.9	6	168	67
85	F	161	47	21.2	5	220	54
86	F	168	53	19.2	3	210	68
87	M	170	66	18.6	7	220	60
88	M	174	67	20.4	7	200	53
89	F	155	49	19.7	1	170	62
90	M	160	52	19.5	8	220	72
91	M	172	64	19.5	9	230	75
92	F	166	58	18.4	8	235	42
93	F	161	64	18.8	5	250	33
94	F	160	51	18.6	2	184	47
95	M	191	96	18.8	6	190	67
96	M	174	52	22.0	2	140	52
97	F	168	52	18.9	4	155	74
98	F	176	70	19.1	1	262	53
99	F	162	49	18.9	4	238	63
100	F	161	49	18.8	4	171	61

C.2 Exercises using student data set

Chapter 3 Sorting and ranking
1 Sort on H or W or A or L or R or C (any sorting method).
2 List the student numbers in order of age.
3 Exercises 3.9, question 9: Run using H or W or R separately for each sex.
4 Exercises 3.9, question 11: Run using any column of the data but separating the sexes.

Chapter 4 Inspection and summary of data using tables
1 Exercises 4.9, question 3: Run using (i) L in classes of unit width; (ii) H in classes of width five.
2 Exercises 4.9, question 5: Run using any column of the data but separating by sex.

3   Modify Algorithm 4.6 (*Two-way Table*) using a classification into sex (M, F) and handedness (L > 10, L ≤ 10). Then run with Algorithm 4.7 to produce the frequency table.

4   Exercises 4.9, question 6: Run with $(x, y) = (H, W)$, separating the sexes.

[*Note*: be careful in choice of class widths.]

5   Exercises 4.9, question 6: Run with $(x, y) = (S, L)$, with L grouped into (0, 5), (6, 10), (11, 15), (16, 20).

Chapter 5   Inspection and summary of data using graphical methods

1   Produce histograms for each of the columns.

2   Exercises 5.10, question 2: Run using each column of the data.

[*Note*: A should be converted to a 3-digit integer.]

3   Exercises 5.10, question 3: Run using each column of the data but separating the sexes, and printing the plot for one sex directly underneath the plot for the other sex.

Chapter 6   Computation of variance and correlation coefficient

1   Calculate and print out the mean and variance of each column of the data, separating the sexes.

2   Calculate and print out the correlation coefficient of the following pairs of variables: (H, W), (W, A), (L, C), (R, C). (Separate the sexes for the first two pairs.)

3   Exercises 6.10, question 7: Run with the pairs of variables suggested in (2) above.

Chapter 10   Significance tests and confidence intervals

1   Carry out a chi-squared test on the contingency table produced above in Exercise 3 for Chapter 4.

2   Exercises 10.7, question 10: Test whether (i) heights, (ii) weights are normally distributed for each sex.

3   Compare (i) reaction times R and (ii) card sorting times C for the two sexes using: (a) *t-Statistic, Unpaired Samples* (Algorithm 10.4); (b) *Mann–Whitney Test*, (Algorithm 10.5).

Bibliography

ABRAMOWITZ, M, and STEGUN, I A, *Handbook of Mathematical Functions*. Dover, New York (1972)

ASHBY, T, A modification to Paulson's approximation to the variance ratio distribution. *The Computer Journal*, 11, 209–10 (1968)

CHAMBERS, J M, Computers in statistics. *American Statistician*, 34, 238–43 (1980)

CHATTERJEE, S, and PRICE, B, *Regression Analysis by Example*. Wiley, New York (1977)

CLARKE, G M, *Statistics and Experimental Design*. 2nd Edition, Edward Arnold, London (1980)

CLARKE, G M, and COOKE, D, *A Basic Course in Statistics*. Edward Arnold, London (1978)

DRAPER, N R, and SMITH, H, *Applied Regression Analysis*. 2nd Edition, Wiley, New York (1981)

FISHMAN, G S, *Principles of Discrete Event Simulation*. Wiley, New York (1978)

GEARY, R C, The frequency distribution of the quotient of two normal variables. *Journal of the Royal Statistical Society*, 93, 442 (1930)

HOGG, R V, and CRAIG, A T, *Introduction to Mathematical Statistics*. 3rd Edition, Macmillan, New York; Collier-Macmillan, London (1970)

JOHNSON, L W, and RIESS, R D, *Numerical Analysis*. Addison-Wesley, Reading, Mass. (1977)

KNUTH, D E, *The Art of Computer Programming*. Vol. 2, Seminumerical Algorithms. 2nd Edition, Addison-Wesley, Reading, Mass. (1981)

KNUTH, D E, *The Art of Computer Programming*. Vol. 3, Sorting and Searching. Addison-Wesley, Reading, Mass. (1973)

LAU, C L, Algorithm AS 147. A simple series for the incomplete gamma integral. *Applied Statistics*, 2, 113–14 (1980)

LORIN, H, *Sorting and Sort Systems*. Addison-Wesley, Reading, Mass. (1975)

MAJUMDER, K L, and BHATTACHARJEE, G P, Algorithm AS 63. The incomplete beta integral. *Applied Statistics*, 22, 409–11 (1973)

MONRO, D M, *Basic BASIC*. Edward Arnold, London (1978)

PAULSON, E, An approximate normalization of the analysis of variance distribution. *Annals of Mathematical Statistics*, 13, 233–5 (1942)

SNEDECOR, G W, and COCHRAN, W G, *Statistical Methods*. 7th Edition, Iowa State University Press, Ames, Iowa (1980)

TOCHER, K D, *The Art of Simulation*. English Universities Press, London (1963)

TUKEY, J W, *Exploratory Data Analysis*. Addison-Wesley, Reading, Mass. (1977)

WALLACE, D L, Bounds on normal approximations to Student's and the chi-squared distributions. *Annals of Mathematical Statistics*, 30, 1121–30 (1959)

WALPOLE, R E, *Introduction to Statistics*. 2nd Edition, Macmillan, New York; Collier-Macmillan, London (1974)

WILKINSON, J H, *Rounding Errors in Algebraic Processes*. HMSO, London (1963)

WILSON, E B, and HILFERTY, M M, The distribution of chi-square. *Proc. National Academy of Sciences*, 17, 684–8, Washington (1931)

YAKOWITZ, S J, *Computational Probability and Simulation*. Addison-Wesley, Reading, Mass. (1977)

Index